Walter Rye

Records and Record Searching

A Guide to the Genealogist and Topographer

Walter Rye

Records and Record Searching
A Guide to the Genealogist and Topographer

ISBN/EAN: 9783337172497

Printed in Europe, USA, Canada, Australia, Japan

Cover: Foto ©berggeist007 / pixelio.de

More available books at **www.hansebooks.com**

RECORDS

AND

RECORD SEARCHING:

A Guide to the Genealogist and Topographer.

BY

WALTER RYE.

LONDON:

ELLIOT STOCK, 62, PATERNOSTER ROW.

NORWICH: A. H. GOOSE AND CO.

1888.

PREFACE.

I FEAR I shall be blamed by many for my temerity in printing what little I have learned during the last quarter of a century about the Public Records,—especially by those who, being better qualified than I to undertake the task, have refrained from doing so, in the hope that some day they may bring out a perfect Handbook to such Records.

That none of my critics *in posse* have taken the trouble to print such a work is my excuse for putting forward this very indifferent substitute for it.

The facts are these : I had, from time to time, collected a mass of notes for my own use, and found that it was very inconvenient to refer to them in MS., so thought that if they were printed and well indexed [1] it might help beginners, and save them some of the trouble and mental worry from which I suffered, when I was first let loose on the enormous mass of the Records in Fetter Lane and Bloomsbury.

---

[1] I am not going to apologise for my Index, for it is the only good part of my book, being adapted to the meanest capacity.

I am indebted to many friends for help in completing this little book, and very especially to my schoolfellow Mr. R. Howlett, to Mr. Walford D. Selby, Dr. Jessopp, Dr. Marshall (Rouge Croix), Mr. C. H. Athill (Blue Mantle), Mr. A. R. Bax, and to others.

That I must have made innumerable omissions and mistakes I know well enough ; but I ask my readers to be merciful, and to send me, more in sorrow than in anger, their corrections and additions. If they will all do so, possibly a second edition may some day be issued, in which my correctors will benefit by their own corrections.

W. R.

WINCHESTER HOUSE,
      PUTNEY, S.W.

CONTENTS.

CHAPTER I.   PAGE

   How to compile a Pedigree   .   .   1

   Specimen Pedigree   10

   Signs, Abbreviations, and Terms used for Pedigrees   11

CHAPTER II.

   How to write the History of a Parish or other Place   12

CHAPTER III.

   Documents relating to the Subinfeudation, Sale, and Transfer of

      Land   21

   Specimens of Pipe Rolls in Tabular Form   .   between 25, 26

CHAPTER IV.

   The Sale and Transfer of Land inter Vivos   .   32

   List of Published Calendars of Fines   .   . 40—42

CHAPTER V.

   Legal Proceedings relating to (1) Land, (2) other Matters not

      criminal   44

CHAPTER VI.

   Of Criminal Proceedings, &c.   51

CHAPTER VII.

   State Papers, &c.   55

CHAPTER VIII.

   Ecclesiastical and Monastic Records, Guilds, &c.   64

CHAPTER IX.

   Parish Registers, Cemetery Books, General Registry Office,

      Churchwardens' Books, Inscriptions, &c.   .   . 73

   List of Published Parish Registers   .   . 76, 77

CHAPTER X.

   Fiscal Records, the Subsidy Rolls, &c.   82

CHAPTER XI.   PAGE
   The Descent of Land, Inquisitions Post Mortem, Proofs of Age,
     Wills and Administrations   85
   Printed Collections of Wills   .   91, 92

CHAPTER XII.
   Manorial Records, Court Rolls, &c.   93

CHAPTER XIII.
   Grants from the Crown, Privileges, Titles, &c.   95

CHAPTER XIV.
   The Record Office   .   .   103
   Record Agents, Transcribers, &c.   .   108
   Chart of the Round Room, Record Office .   112
   Chart of the Long Room   ,,   ,.   114
   The British Museum Reading Room Chart   .   116
     ,,   ,,   ,,   Rules and Regulations   118
   Probate Registry   .   121
   City of London Records   125
   Lambeth Library   125
   The Heralds' College   .   129
   Printed Visitations   .   132
   The Bodleian Library .   136
   Cambridge University Library   137

Antiquarian Booksellers, &c.   .   138

APPENDIX I.—Form of Writ of Diem Clausit Extremum   139
           Form of Inquisition Post Mortem   139
           Form of Testament   141
           Probate Act   .   141
           Specimen of Nuncupative Will .   142
           Form of Fine   .   .   142
           Form of Charter   .   .   144

APPENDIX II.—Registrar General's Documents   146

APPENDIX III.—List of London Cemeteries   .   .   148

APPENDIX IV.—Classified List of the Master of the Rolls Series of
       Chronicles, &c.   150

APPENDIX V.—List of Published Domestic State Papers   .   157

APPENDIX VI.—Classified List of the Historical MSS. Commission,
       Reports (1) Places   .   160
             (2) Persons   .   163

APPENDIX VII.—A short Antiquarian Directory   166

Index   .   172

CHAPTER I.

How to compile a Pedigree.[1]

In working up a pedigree you should always begin with the earliest *undoubted* fact in your possession.

Aged relations and friends, inscriptions in bibles, prayer books, &c.; deeds, probates, old letters, and tombstones, are among the most obvious sources of information: and what is obtained from them had better be worked into a tabular form before any search is made in the register of the parish in which your family resided, as there is often considerable difficulty in making out a sketch pedigree from the registers, if the entries are numerous, and there were two or more fathers of the same Christian and surname having families at the same time.

Great care should be taken in making notes of the *vivâ voce* evidence of aged people: the result should be read over and over again to them at as long intervals as possible, and presented to them in a different shape each time.

A name or a date which your informant is utterly unable to remember one day will often come to mind after a week or so; and, moreover, by the manner in which your informant adheres to or varies former statements, you can form a tolerably accurate idea as to how much you can rely on them.

---

[1] A recent work which should be consulted is *How to Write the History of a Family*, by W. P. W. Phillimore (Elliot Stock.) When I mention that it is revised as to various branches by such experts as Mr. Selby of the Record Office, Mr. Smith of the Probate Office, and Mr. Cokayne (Norroy King-at-Arms,) I need hardly say that it will be most valuable. It was not issued when this chapter was written, and I have purposely abstained from looking at it, except to work some references to it into my index.

Note everything relating to change of residence, profession or occupation, relationship, or connections however distant; for though while you are doing so your notes may seem utterly worthless, they may prove of great value to you hereafter in tracking out a clue or eking out a coincidence.

Sooner or later, however, perhaps at your grandfather or your great grandfather, or perhaps a step or two higher, you will come to the end of your oral, and also of your documentary evidence, and the head of your sketch pedigree (vide specimen pedigree) is Henry Smith, who died at Fulham in 1768, aged 75.[1]

It is most likely some of your relations know whence he originally came, if he was not a native of the place in which he died. In that case search should be made in the parish register[2] of his reputed birthplace for the baptism of a Henry Smith, in or about the year 1693. You can never depend on the age being right within a year or two.

If you find this entry it will give you the Christian names of his father and mother, and carry you a step higher; and you should then work the pedigree up by the registers of the same parish as long as there are any entries of persons of your name. Keep on working out the results in a tabular form.

Possibly, however, you have been mistaken or wrongly informed as to the parish in which he was born, or though born he may not have been baptized in it, in which case you should try the adjacent churches, and failing in them it may be as well for you to search the registers of any parishes in which any of your family, or connections by marriage, were formerly resident, even for a short time.

If still unsuccessful you will be in the same predicament as

[1] If you cannot find his age on his tombstone or burial certificate you may do so in any policy on his life, which also frequently has appended to it a copy of his baptismal certificate produced by him to show his age when he insured his life.

[2] If not a member of the Church of England, see notices of dissenting registers, p. 79.

many who continually advertize [1] in *The Times* and elsewhere, offering liberal rewards for copies of baptismal certificates of people known to have been born in certain specified years.[2]

Nominally, the way out of your difficulty is easy; but practically, it is very hard. There *should* be at the different Bishop's Registries transcripts of all parish registers of the dioceses sent in year by year, and called Register Bills; and knowing the year, nothing theoretically is easier than to search the bundle for such year, and find your entry in its proper parish. But, unhappily, this cannot be depended on at all. The great majority of the clergy and churchwardens have been most remiss in sending in these duplicates, and owing to this and the little store set on them by the authorities, who should be

---

[1] While touching on advertisements let me warn you against the catchpenny " next of kin " advertisements, by which rogues try to rob poor people with specious tales of unclaimed stock. Most of the statements they contain are absolute lies; *e.g.*, that which one reads every day, about there being above ten millions of unclaimed property in the Court of Chancery ; the truth being that there is next to none ; and I can only say that in the experience of my late father and myself, which covers sixty years, not a single case has occurred in which anyone has benefited in the least by any advertisement of this sort. I have myself investigated many instances in which what seemed most plausible *primâ facie* cases have been put before me at first, and have invariably found that the claimants have been " kept out of their property " by—the rightful owners! The following advertisement, to which I will give a gratuitous circulation, is a fine example of the simple faith of a claimant :—" Required, the address of the Solicitor that has the property of " one believed to be John Marsh, who died about one hundred years ago. This " property supposed to be in Manchester and other places. The Marsh family, " of Purbeck in Dorset, believed to be the heirs. This property supposed to " be in Chancery. If he will not send his address, will he say in what year " the will of this property was made, and in what county?—Arthur Marsh, 34, " Floating Bridge Cottage, Southampton."

[2] There is a Society of Parish Clerks—Parish Clerk's Hall, E.C.—who profess, for a fixed fee of £2. 2*s.*, to have all the registers within the Bills of Mortality searched, provided a reward is also offered for the certificate. This search is not often a fruitful one. If your ancestors held a good position it may be worth your while to search the births, deaths, and marriages in the *Gentleman's Magazine* and the *Annual Register*. The former has Indexes Nominum for each year; the latter has a general Index from 1758 to 1819. See also Chapter IX., on Registers, *sub voce* " Dissenters."

4

their custodians, but a very small percentage now exist. Still they are always worth searching, if only for the off-chance, and they have stood me in good stead more than once. The fee usually charged is 3s. 4d. a year.

As with your oral evidence, however, so with your parish register; you must get to the end of it sooner or later. These registers were instituted in 1538, but only very few begin as early as this, and 1680 may be taken as a fair average, though, unluckily, many begin still later.[1]

But before quitting parochial records altogether, it may be as well to enquire for any parish or town books giving accounts of rates collected from inhabitants, as they will sometimes, though of course but imperfectly, serve to fill up a gap in the parish register. In cities the Freemen's Roll is most useful, and is in the custody of the Town Clerk. They are especially useful, moreover, to prove the residence of anyone in a parish, in which he may not have either been born or married, or have died.

The printed Poll Books, of which there is a fine series in the British Museum, begin at the early part of the last century, and are often very useful as giving the searcher a hint as to where persons of his surname were settled in the country.

American searchers will find Hotten's *Lists of Emigrants* (1600-1700), published by Chatto and Windus, 1874, very useful, as they will the *Licenses to pass beyond Seas*, &c. (see 20th Report, p. 131); but are specially warned against all advertizing record agents, for rich Americans, in their anxiety for pedigrees, are lavish in payment, and are therefore the special marks of certain unscrupulous individuals.

If the family about which you are searching were of any importance, the *Gentleman's Magazine* (1731) and *Annual Register* (1758) may be searched for obituary notices, marriages, births, &c. The Index Society has just published an excellent[2] Index

---

[1] The dates at which the registers of any parish begin may be ascertained by the abstract of replies, referred to in Chapter IX. p. 35, under the title of *The Parish Register Abstract of* 1830, which was printed by order of the House of Commons in 1833.

[2] Though it has been spitefully reviewed.

to the Obituary and Biographical Notices in the *Gentleman's Magazine*, from 1731 to 1780, which of course supersedes the old faulty indexes between the dates. It also proposes to publish Annual Indexes of Obituary Notices. That for 1880 has been issued.

If you can obtain access to the Court Rolls of any manors[1] which may exist in the parish, they will often be of the greatest possible value to you.

Some information may also be obtained from the local antiquary, or failing him, from the local solicitor, who is generally not indisposed to help you, if he can do so without disclosing the title of any client.

There is one more class of records, divided into localities, to which I had better refer before touching on the general sources of information, and that is the Subsidy Rolls,[2] which will give you the names of men taxed in different villages all over England in early times.

Your next and most productive field of labour lies among the Wills at the different Registries.[3]

You should first search in your own District Registry, and after referring to and taking notes of the wills of all your own name, who you know to have been of your own family or resident in your locality, proceed to do the same with all of your name residing in other places.

Work out the results of each will, in a tabular form, on a separate sheet of paper, and put it by till you have done all, and then spread out the lot before you, and try to connect them.

Never attempt to theorize or speculate too soon. If you get wedded to a theory, you will find yourself unconsciously specially pleading in its favour, and not looking at things fairly.

Your District Registry exhausted, try the Principal Registry in London, in which are the wills of many of the more important country folks.

You have now come to the end of all your MS. and local

---

[1] Vide p. 93.　　　　　　[2] Vide p. 83.

[3] There are very many collections of wills, printed and in MS., referred to more at length on p. 91.

records, and I most strongly advise you never to look at a printed book until you have done so.

When you have done this, you may search the indexes of all county histories and other topographical[1] and genealogical[2] works, both printed and MS., relating to your locality, which are to be found in the museums and libraries mentioned in Chapter XIV., and then the Heralds' College, though, if you are working up a new pedigree of a non-armigerous family, you will get little there.[3]

On the earlier descents drawn up by Heralds place but very little reliance; but they may be referred to, and you will find an excellent list of the visitations by Dr. Marshall in the *Genealogist*, new series, i. p. 201, and a digest of what have been printed at p. 132, *post*. There are also lists by Sims and in the two Guides. The indexes to these printed Heralds' visitations (see Chapter XIV. p. 132), to the printed parish registers (see Chapter IX.), to the *Collectanea Topographica et Genealogica*, and its numerous successors, and to the old edition of Dugdale's *Monasticon*, should then be searched.

Then the mass of available printed matter open to you must be attacked. There is much information as to the matches, births, and deaths of armigerous families to be found in Peter Le Neve's *Memoranda on Heraldry* (1695-1729), which is printed in the *Topogr. and Geneal.*, ii. 25 et seq. Mawson's Obits, &c. (1720-1729), now being printed in the *Genealogist*, new series, ii. and iii.; and Le Neve's Catalogue of Knights, edited by Dr. Marshall for the Harleian Society, with further additions thereto, in the *Genealogist*, 1st series, i., should also be searched. Sir William

---

[1] For these see Anderson's very valuable *Book of English Topography*.

[2] All English pedigrees of more than three generations are admirably indexed in Dr. Marshall's *Index to Printed Pedigrees*, which is now in its second edition. This includes references to all pedigrees in Burke's *Landed Gentry*, and other books of the same class.

[3] I need hardly tell you to avoid, like poison, all advertisements offering to send you, in return for your name and county and fee, "a plain or coloured" sketch of your arms. The "Heraldic Office" stock-in-trade is usually only a copy of Burke's *Armory*. You can get the same information yourself without payment.

Musgrave's Obituary of the Nobility and Gentry was begun (some little time ago) to be reprinted by Mr. Foster, vol. i. of his series styled *Collectanea Genealogica*, but vols. ii. and iii. had no more of it, and parts xiv.-xx. were taken up by a modern law list, called "Men at the Bar," so I, for one, left off subscribing to the series.

Then the Marriage Licenses must not be forgotten, for they give very valuable information as to the parentages and ages of the persons named in them.

The London Marriage Licenses (including Faculty, Westminster, &c.), 1521-1869, are printed in one vol. by Mr. Joseph Foster, from the original MS. collections of the late Col. Chester; and the same collections, but in a less consultable form, were printed by the Harleian Society, and issued to their members.

Some extracts from the Marriage Licenses in the Diocesan Registry at Worcester, are printed in the *Genealogist*, 1st series, vii. p. 262; 2nd series, i. pp. 27, 102, and ii. pp. 52, 150, 212.

Mr. Gibbons is going to print the Allegation Books for Lincoln, from 1598 (Mitchell and Hughes).

Funeral Certificates, given by the Heralds, often contain most valuable information, for they are certified by the son and heir or other near relation of the deceased, and give his marriage, family, &c. For specimens of them see *Genealogist*, new series, i. p. 186, ii. p. 85; the volume of Lancashire Funeral Certificates (1600-1678), edited by Rylands (Record Society), 1882; and King's volumes of Funeral Certificates (Chetham Society). Of course, too, books of Monumental Inscriptions, like Le Neve's *Monumenta Anglicana* and Dingley's *History from Marble*, will not be forgotten.[1]

Other sources worth searching are the Reports of the Historical MSS. Commission (which are well indexed) and the Calendars of Domestic State Papers,[2] while, if the person searched after was a clergyman, Cooper's *Athenæ Cantab.* (1500-1609), Wood's *Athenæ Oxonienses*, Foster's *Oxford Matriculations* (1581-1886), printed from Col. Chester's MSS., now being issued, and his *Alumni Oxonienses* (1715-1886), or the *Register of the University*

---

[1] For early London inscriptions see the "Catalogue" of those before the fire of 1666, by Fisher and Morgan.

[2] For lists of these to date see Appendix V., p. 157.

of Oxford. now published by the Oxford Historical Society (1449-1571 published), may be searched. For nearly every public school a book containing a list of old scholars, often giving the father's name, has been published, *e.g.*, Harwood's *Alumni Etonensis* (1443-1797), Welch's *Queen's Scholars of Westminster* (1788-1852). Robinson's *Merchant Taylors*, Trollope's *Christ's Hospital*, Haig-Brown's *Charterhouse*, Parish's *Carthusians* (1800-1879), Gardiner's *St. Paul's* (1512-1876), Forshall's *Westminster School*, and the *Rugby Register* (1675 to the present time.)

Should you have reason to suppose your ancestor was a doctor, consult Munk's *Roll of the Royal College of Physicians* (1882); a lawyer,[1] *The Law List*, which begins 1775; a literary man, Watts' *Bibliotheca Britannica* (4 vols. 4to. 1824), or the last edition of Lowndes' *Bibl. Manual;* or a politician, Beatson's *Political Index*, or his *Chronological Register*.

Last, but not least, there will remain to you the contents of the Record Office—a mine of material so rich that you could not search a tithe of it if you were to spend a hundred years there.

Perhaps the most productive, or rather the easiest searched documents there, are the Parliamentary Surveys, Royalist Composition Papers, Feets of Fines, State Papers, and Chancery Proceedings, to which you will find reference in the index.[2]

But the general contents, the De Banco Rolls, the Coram Rege Rolls, &c., are from their immense bulk and chaotic arrangement utterly sealed to you, unless you by chance get a reference to the roll and skin. The frequently inserted advertisement that a certain record agent has an "index"

[1] The registers of the different Inns of Courts should also be searched, but are difficult of access. Mr. J. Foster has issued, or is about to issue, a *Hand List of Admissions to Lincoln's Inn*, 1420-1885, and proposes to issue similar lists of the other Inns.

[2] Unless you wish to know "the truth, the whole truth, and nothing but the truth" about your family, do not too closely search the Crown Plea Rolls, or you may_feel disgusted at finding that an ancestor was hanged for murder, burglary, or some other trifle.

to these De Banco Rolls is most misleading. It is absolutely impossible that any one man could index a tithe of them in a long life. He can have an index to his notes or extracts only.

Here I must leave you with the advice, if you wish to make out a perfect and ample pedigree of your family, to undertake the history of the district or parish with which they had most to do.

Things then turn up in most unlikely corners, in records which you would never have the pluck to specially search for your name; the meshes of the net of your search grow smaller and smaller, and all the while you are doing good archæological work.

One more last word on this subject. It is more honourable by far to trace your descent sturdily and clearly to some mediæval yeoman or tradesman. than by straining coincidences, presuming identities, and fudging judiciously, to attempt to hook on to some noble[1] or well-descended family of the same name as yours, or something like it, e.g., the recent instance of De Bréci and Brassey. If you want a sad and degrading instance of a highly gifted man and clever antiquary giving way to this latter weakness, read " The Chandos Peerage Case " by Beltz, and remember there are still plenty of Beltzes ready to cut you up root and branch, as severely and trenchantly as was done in this book.[2]

---

[1] If you have reason to suspect you have a "royal descent" (as in all probability you must have if you have any gentle blood in you at all), you had better consult Burke's *Royal Descents* and Foster's *Royal Lineage*, of which there are several series.

[2] Place little reliance on the pedigrees in most Peerages, and books treating on the pedigrees of the well-to-do classes. The editors have, in too many cases, inserted concocted pedigrees, sent them by interested persons. It is next to impossible to check every statement, e.g., I heard the other day of a case where notoriously illegitimate issue were inserted as legitimate. Cokayne ("G. E. C."), Doyle, and Foster, are honourable exceptions to the general rule. *Popular Genealogists; or, the Art of Pedigree Making*, Edinburgh, 1865, is well worth perusal.

Specimen Pedigree.

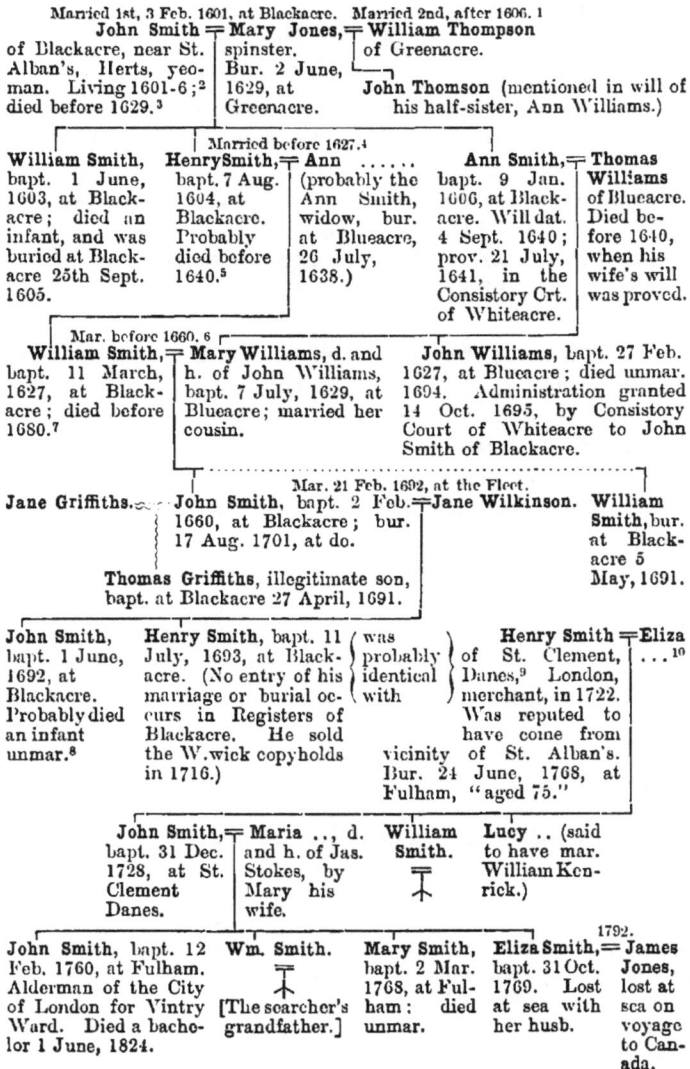

Married 1st, 3 Feb. 1601, at Blackacre. | Married 2nd, after 1606. 1

John Smith ⊤ **Mary Jones,** ⊤ **William Thompson**
of Blackacre, near St. | spinster. | of Greenacre.
Alban's, Herts, yeo- | Bur. 2 June, |
man. Living 1601-6;[2] | 1629, at | **John Thomson** (mentioned in will of
died before 1629.[3] | Greenacre. | his half-sister, Ann Williams.)

Married before 1627.[4]

William Smith, | **Henry Smith,** ⊤ **Ann** | ...... | **Ann Smith,** ⊤ **Thomas**
bapt. 1 June, | bapt. 7 Aug. | (probably the | bapt. 9 Jan. | **Williams**
1603, at Black- | 1604, at | Ann Smith, | 1606, at Black- | of Blueacre.
acre; died an | Blackacre. | widow, bur. | acre. Will dat. | Died be-
infant, and was | Probably | at Blueacre, | 4 Sept. 1640; | fore 1640,
buried at Black- | died before | 26 July, | prov. 21 July, | when his
acre 25th Sept. | 1640.[5] | 1638.) | 1641, in the | wife's will
1605. | | | Consistory Crt. | was proved.
| | | of Whiteacre.

Mar. before 1660. 6

William Smith, ⊤ **Mary Williams,** d. and | **John Williams,** bapt. 27 Feb.
bapt. 11 March, | h. of John Williams, | 1627, at Blueacre; died unmar.
1627, at Black- | bapt. 7 July, 1629, at | 1694. Administration granted
acre; died before | Blueacre; married her | 14 Oct. 1695, by Consistory
1680.[7] | cousin. | Court of Whiteacre to John
| | Smith of Blackacre.

Mar. 21 Feb. 1692, at the Fleet.

Jane Griffiths. ⸱⸱⸱ **John Smith,** bapt. 2 Feb. ⊤ **Jane Wilkinson.** | **William**
| 1660, at Blackacre; bur. | **Smith,** bur.
| 17 Aug. 1701, at do. | at Black-
| | acre 5
Thomas Griffiths, illegitimate son, | May, 1691.
bapt. at Blackacre 27 April, 1691.

John Smith, | **Henry Smith,** bapt. 11 | (was | **Henry Smith** ⊤ **Eliza**
bapt. 1 June, | July, 1693, at Black- |) probably | of St. Clement, | ... [10]
1692, at | acre. (No entry of his |) identical | Danes,[9] London,
Blackacre. | marriage or burial oc- | (with | merchant, in 1722.
Probably died | curs in Registers of | | Was reputed to
an infant | Blackacre. He sold | | have come from
unmar.[8] | the W.wick copyholds | vicinity of St. Alban's.
| in 1716.) | Bur. 24 June, 1768, at
| | Fulham, "aged 75."

John Smith, ⊤ **Maria** .., d. | **William** | **Lucy** .. (said
bapt. 31 Dec. | and h. of Jas. | **Smith.** | to have mar.
1728, at St. | Stokes, by | ⊤ | William Ken-
Clement | Mary his | ⊥ | rick.)
Danes. | wife.

1792.

John Smith, bapt. 12 | **Wm. Smith.** | **Mary Smith,** | **Eliza Smith,** ⚌ **James**
Feb. 1760, at Fulham. | ⊤ | bapt. 2 Mar. | bapt. 31 Oct. | **Jones,**
Alderman of the City | ⊥ | 1768, at Ful- | 1769. Lost | lost at
of London for Vintry | [The searcher's | ham: died | at sea with | sea on
Ward. Died a bache- | grandfather.] | unmar. | her husb. | voyage
lor 1 June, 1824. | | | | to Can-
| | | | ada.

For notes see next page.

1 That being the date of the birth of her last child by her first husband.

2 In which years he had children baptised there.

3 The date of the death of his widow.

4 That being date of baptism of his first child.

5 As he is not mentioned in will of his sister Ann Williams of that date.

6 The date of the baptism of his son.

7 When his son Henry was admitted as his heir to his copyholds of the manor of W.wick in Blackacre.

8 If he had not died unmarried, and without issue, his younger brother Henry could not have sold the W.wick copyholds in 1716.

9 Occurs so described in the rate books of this date.

10 Her maiden name said to be Johnson.

--- ---

Signs, Abbreviations, and Terms used for Pedigrees.

= (or joined hands)	.	married to.
ob. . . .	.	died.
ob. inn. . .	.	died unmarried.
ob. cœlebs .	.	died a bachelor.
ob. s. p. .	.	died without issue.
ob. s. p. m.	.	died without male issue.
ob v. p. .	.	died in the life of his or her father.
⊤⼂ . .	.	had lawful issue not specified on pedigree.

[The use of these abbreviated Latin terms is now not so common as it formerly was. Facts are now generally stated at length, or reasonably abbreviated as in the specimen pedigree opposite.]

s. and h. . .	.	son and heir.
d. and h. . .	.	daughter and heiress.
d. and coh. . .	.	daughter and coheiress.
s. and coh. .	.	sister and coheiress.
....	.	probable relationship.
~~ .	.	illegitimate relationship.

CHAPTER II.

𝕳𝖔𝖜 𝖙𝖔 𝖜𝖗𝖎𝖙𝖊 𝖙𝖍𝖊 𝕳𝖎𝖘𝖙𝖔𝖗𝖞 𝖔𝖋 𝖆 𝕻𝖆𝖗𝖎𝖘𝖍 𝖔𝖗 𝖔𝖙𝖍𝖊𝖗 𝖕𝖑𝖆𝖈𝖊.[1]

PERHAPS the best thing to begin with is the Church, which will probably be the oldest edifice in the parish or place whose history you have undertaken to write; and certainly has interwoven with its history that of most of the families in the parish.

First copy all the Inscriptions within its walls, take rubbings of the brasses, if any, and note all the heraldry both in the windows and on hatchments, but you need not copy *verbatim* the later inscriptions in the churchyard, or at all events you can omit the poetry which embellishes them.[2]

Note as many of the Entries in the Parish Registers[3] as you can, especially for the earlier years, and at once index up the notes you get from them and your visitation of the church.

Churchwardens'[4] and Town Books (which will generally be found in the church chest) should be dealt with in the same way, and you should look in them for entries relating to repairs done to the church, the sale of church ornaments, &c. They will also sometimes be useful for the lists they give of parishioners paying rates or receiving parochial relief during periods for which the parish registers may be imperfect.

[1] It may be as well for me to say that this title headed this chapter when I submitted the MS. to the late Mr. J. G. Nichols many years ago, so it is not plagiarized from the Rev. D. Cox's work of the same title, which I have never seen, though it is spoken of most highly, and has now reached a third edition.

[2] I always note the *facts* only which they contain.

[3] As to obtaining access to them, fees, &c., see Chapter IX.; see also the list of printed registers given by Dr. Marshall in *Geneal.*, 2nd series, ii. pp. 194-201, as added to on p. 77.

[4] As examples of a printed Churchwardens' Book, I may cite that of St. Martin's, Leicester, 1489-1884, which has recently been published by T. North, F.S.A., and Bax's *Early Churchwardens' Accounts of Hurley* (Surrey Arch. Soc.)

Having gone so far, transcribe on separate half sheets of foolscap paper (writing on one side only, with a wide margin) any printed accounts of the parish which you can find in topographical works relating to the county or district, keeping them as much as possible divided into classes, such as (*a*) the Church and its Rectors or Vicars, (*b*) the Manor or Manors, (*c*) Miscellaneous History.

Note on separate slips of paper every reference they give to original documents, but keep on indexing. In short, never leave an item of newly acquired information unindexed for a day.

Then go through and check such references with the original documents in the Record Office, Probate Courts, the Copyhold, Enclosure, and Tithe Commutation Offices; the Ecclesiastical Commission, the Land Revenue Record Office, and elsewhere, as far as you possibly can, and if practicable, transcribe each document at length, always on separate sheets of loose paper, and never in a book, and you will always then be able to arrange your documents in order of date or in any other way you wish, and you will never be embarrassed by the arrival of fresh matter. Clips like music books with springs at the back are sold by most stationers, and are better than a loose portfolio.

Initial every statement in your printed authority as you verify it, and take it as a golden rule never to trust to anyone's work but your own.[1]

Having worked up all your references, begin on your own account to search for more.

You cannot begin better than by trying *Kemble's Codex*, the *Anglo-Saxon* and all other published *Chronicles*, in the Rolls,

---

[1] The most careful topographers are liable to mistake, *e.g.*, Blomefield, in his "*History of Norfolk*" *sub voce* Cromer, says that certain land in Shipden was granted to the Prior and Convent of the Carthusians by Sir William Beauchamp, and refers to the Patent Roll, 16 Rich. II. When I was writing the History of this Church this puzzled me amazingly, for I knew nothing of any Beauchamps connected with the place. On referring to the roll, I found that the land was given by one Geoffrey de Somerton, and that the mistake had arisen from the license in mortmain, obtained by the Carthusians, including some land in London given them by Sir William Beauchamp.

and other series, the old edition of the *Monasticon*, the *Calendar of Bodleian Charters*, &c., for any incidental mention of your village in early times. The *Monumenta Historica Britannica*, by Petrie and Sharp, 1 vol. fo. (1848), which sells for about £2. 15s., should be consulted for the Saxon period.

Domesday will be tolerably sure to give you some information, nor should you neglect the local so-called *Domesdays* of Exon, Winton, St. Paul's, Norwich, &c.[1]

By all means, if possible, obtain access to the early Court Rolls of any manors in your parish,[2] which will afford you a great deal of interesting information as to the manners and customs of the tenants, and names of localities and fields now forgotten or corrupted, and also very many particulars of the descent of families.

From the heads of the rolls which give the names of the lords holding the different courts you can make out a skeleton descent of the manor or manors, which you can then fill in from the inquisitions post mortem[3] and wills[4] of the lords of the manor, the monuments in the church, the De Banco Rolls, Feet of Fines, and other documents relating to the transfer of land contained in the Record Office, and enumerated in my third chapter.

Similar descents should be made, from the same sources, of all the most considerable estates.

Of the more important families who have ever lived in your parish you should draw up pedigrees from Church Registers, Wills, Monuments, Inquisitions Post Mortem, &c., and you should see if there are any such already drawn out in the *Heralds' Visitations*[5] or *County Histories*, and if so you should copy and correct them.

In aid of these pedigrees, and also to obtain a just idea of

[1] See Chapter III., p. 21.

[2] How to obtain access to them, see p. 93. For forms, see Appendix I., p. 139. There are many court rolls in the Record Office once belonging to the Augmentation Office. See 20th Report, p. 80, and 45th Report. A great number were also transferred with the Duchy of Lancaster Records; see list in 43rd Report, pp. 206-362.

[3] See Chapter XI. p. 85.   [4] See Chapter XI. p. 90.
See Chapter XIV. p. 132.

the number and importance of the inhabitants of your village in former times, you should search the *Lay Subsidy Rolls*,[1] from the assessments in which you will see who were the more important inhabitants, and this will give you a clue in what names to search for wills, &c., for the wills of the richer inhabitants not only give you information about their families, but from containing legacies for the repairs of the church, for vestments, tombs, windows, brasses, &c., &c., will often help you in working out the account of the fabric of your church.

The *Taxatio Ecclesiastica*[2] (which is an account of the taxation of the tenths of all ecclesiastical benefices taken about the year 1291); the *Inquisitiones Nonarum*[3] (which were inquisitions taken to assess a subsidy in 1342); and the *Valor Ecclesiasticus*[4] (which was a survey of all ecclesiastical property taken just before the Reformation), are all worth searching, and being printed with good indexes will give you little trouble.

Rough accounts should be drawn up of the descent of the advowson, and of the succession of the rectors or vicars.[5]

The former you will be able to collect from much about the same sources as the descent of the manor (to which it very often belonged), while the latter you will get from the Institution Books at the Registry of the Bishop in whose diocese you may be, and also to a certain extent from the Institution Books at the Record Office, where you should also search the *Clerical Subsidies*.[6] If the see were ever vacant for any time, you should consult the Archbishops' Registers at Lambeth or York for the interval.

If in the Home District, you will find that Newcourt's *Repertorium Ecclesiasticum Londinense*, which contains good indexes of incumbents, will help you.

---

[1] See p. 83.

[2] Printed by the Commissioners of Public Records, 1802, and commonly called the Taxation of Pope Nicholas.

[3] Id., 1807.

[4] Id., 1810, *et seq.*

[5] *The Institutions for Wiltshire*, from 1297 to 1810, were printed at the Middle Hill Press, by Sir Thomas Phillipps in 1825, in folio. It is a rare book.

[6] They date from the reign of Edward I. See index.

Details of the lives and connexions of your clergy you may obtain from their wills and monuments, the parish registers, and also from Wood's *Athenæ Oxon.*, Foster's *Calendar of Oxford Graduates,* and his *Oxford Matriculations,* and Cooper's *Athenæ Cantab.*, and you may as well, too, consult Lowndes' *Bibliogr. Manual,* as the clergy were ever a literary class.

Master's *History of Corpus Christi College, Cambridge,* gives the lives of many Norfolk parsons.

Before you begin to write the history of the church itself you should see if there are any MS. collections relating to it, or old notes of its heraldry or monuments preserved in the British Museum or elsewhere. Of course you will take care to go into the MS. Department of the British Museum, and see the great "classed catalogues," which the attendants will show you on request. Should there be any Chantry, you had better consult the Chantry Certificates in the Augmentation Office, taking as a text book Canon Raine's *History of the Chantries within the County Palatine of Lancaster,* 2 vols. (Chetham Society), 1826.[1]

If your parish is in the Eastern Counties, or in Middlesex, you may find a certificate returned into Chancery in 12 Richard II.,[2] by some guild held in your parish. Toulmin Smith's *English Gilds* (Early English Text Society) may be consulted as to these.

In any case you will be tolerably sure of finding a certificate of the sale or an inventory of church goods made or taken in 2 or 6 Edward VI.[3]

The former are especially interesting to the topographer, as

[1] Mr. E. Green is editing a work for the Somerset Record Society on the Chantries of his county. See later as to Chantries, p. 69.

[2] See p. 70. You will also find many references to guilds, lights, &c., in the wills of early inhabitants.

[3] For specimens of these and of a later species of these Inventories see Peacock's *Church Goods of Lincolnshire;* Rye's *Church Goods of Norfolk, Trans. of Norfolk and Norwich Archæological Society,* vol. vii. p. 20; L'Estrange's *Church Goods of St. Andrew and St. Mary Coslany, Norwich, Id.* p. 45; Tyssen's *Inventories of the Goods and Ornaments of Surrey Churches* (1869); Bailey's *Inventories of Lancashire* (Chetham Society), 1879 and Money's *Berkshire Church Goods.*

they give accounts of how the proceeds were spent, generally for re-edifying the church, and often contain entries accounting for the destruction of the rood loft, &c., &c.

Duplicates of these will sometimes be found at the Bishop's Registries.

Other and later Inventories of Church Furniture and Plate are to be found at the Public Record Office, for the reign of Queen Mary, and at the Bishops' Registries for later reigns.

The Terriers, which give accounts of the possessions of the church in each parish, should be transcribed. One copy should be in the church chest, and another in the Bishop's Registry.

The Visitation Books at the Bishops' Registries, with their curious presentments of irregularities, such as of clergy carting dung, and so on, will be most valuable to you, if you can obtain access to them without having to pay the usual heavy fees.

Should there have been any *Abbey* or other Monastic Establishment[1] in your parish you must, of course, see what accounts of it have been already printed in Dugdale's *Monasticon* and elsewhere;[2] ascertain if possible where its chronicle, chartulary, and ledger books are, and do the best you can to digest them.

There may be a " compotus " or account of steward, receiver, or bursar,[3] giving minute and interesting particulars of rents and expenditure, often specifying the daily diet of the inmates.

In any case there will probably be something relating to it in the *Valor Ecclesiasticus*,[4] before referred to ; in the Acknowledgments of Royal Supremacy (taken 1534), the Deeds of Surrender (same period), and the Particulars for Grants, temp. Henry VIII. The last-named series of documents often give detailed descriptions of the state of the buildings, &c.; and you should also refer to the " Letters relating to the Suppression of the Monasteries," published by the Camden Society.

Should you wish to make out a list of all its possessions you had better go through the files of the Feet of Fines, or the

---

[1] As to all of this see more fully Chapter VIII.   [2] Vide page 31.

[3] For reference to printed accounts with notes see page 69.

[4] See Chapter III.

county in which it is situate, search the Fine Rolls for entries of fines paid for licenses to alienate, licenses in mortmain, &c., and try to find a confirmation on the Patent Roll, as the latter will recite the previous gifts. So with the first Minister's Account after the dissolution ; while, for the subsequent disposal of the property, the Particulars for Grants should be looked at.

If there were any King's *Castle*[1] or other building, you should try for mention of it in the Ministers' Accounts, and on the part of the Pipe Rolls which relate to your county. If a private Castle or old Mansion House, you will probably find the license to fortify or "crenellate" it on the Patent Roll, where all such licenses were enrolled.

Licenses for free warren or chase,[2] or to hold a market or fair, you will also find on the Charter or Patent Roll.

As to *Royal Forests* you will find an immense mass of proceedings at the Record Office, and so you will about Fisheries. Some references as to forests will be found in the index, and a Calendar of Proceedings as to Royal Forests (5144 membranes), Henry VIII.—Charles I., will be found in the 5th Report, pp. 48-59.[3]

---

[1] Clarke's *Mediæval Military Architecture in England*, 2 vols. 8vo., 1884, is a useful book on this subject, and Viollet le Duc's *History of a Fortress* may be consulted, as may also be Bond's *Corfe Castle*. There are many rolls as to the repairs of English castles, *e.g.*, Northampton, Winchester, Marlborough, and Somerset, see 1837 Report, pp. 181 and 182 a ; also account as to work done to eleven castles, temp. Edward I., see *Id.*, pp. 185 b. and 191 a ; see also 1819 Report, p. 196.

[2] See *Some Account of English Deer Parks*, by Evelyn P. Shirley, 1867.

[3] These references may also prove useful : Amerciaments of the Forests of Northumberland, 24 Henry III., 1880 Report, p. 205 a ; Copies of Forest Rolls for Oxford, 40 Henry III., *Hales' MSS.*, Lincoln's Inn, lxxvii. No. 9 ; Forest Rolls, 1208-1377, 1800 Report, p. 39 a ; Waste in Divers Forests, temp. Edward I., 1837 Report, p. 13 b ; Account of Forester of Isle of Wight, 22-27 Edward I., 1800 Report, p. 205 b ; Black Book of the Northern Forests, temp. Henry VIII., 1837 Report, p. 12 b ; Charges, &c., in Waltham Forest Courts, temp. Charles I., *Cambridge University MSS.*, D.d. vi. 36 ; Perambulations of Forests, 17 Charles I., 1837 Report, p. 118 a ; Bundles of Claims as to Liberties in, temp. Charles I., 1837 Report, p. 69 b ; Writs for electing Verderers and Regarders of, from 9 Charles I., *id.* p. 118 a ; Survey

If your parish forms part of a town or city a separate class of records altogether will spring up, at which space will only allow me to hint, such as Court and Minute Books, Freemen Rolls, Coroners' Rolls,[1] Letter Books; but the various works published by the City of London, such as *The Liber Albus, The Liber Custumarum,* and so on, will serve as guides to you, as will *The Records of Nottingham* (1155-1547), 3 vols. 8vo., Quaritch; and *The Records of Oxford* (1509-1583), 1 vol. 8vo., Parker.

In writing an account of a city, Madox's *Firma Burgi*, Riley's *Liber Albus of the City of London*, and the various excellent volumes of *The Records of Oxford and Nottingham*, just mentioned, will prove excellent precedents and guides; while Gomme's useful *Index to Municipal Offices* (Index Society), should not be neglected.

The records relating to *Enclosures of Common Lands*,[2] and awards in respect of them before 1810 (?), *ought* to be preserved by the Clerks of the Peace of the different counties; and after that date are to be found at the Tithe Commutation Office, 3, St. James' Square, London, where also is the *Tithe Map*, with its valuable list of field names. A duplicate of the last should be with the churchwarden of each parish. If the new large scale Ordnance or field map of the parish is issued it should be obtained, as it is very valuable to show watercourses, old banks, &c.

Of course, if your county is within the ambit of such a society as the Chetham, the Lancashire and Cheshire, the Middlesex Record Society, the Somerset Record Society, or the Surtees,

of Enfield Chase, 1656, *Cambridge University MSS.*, D.d. ix. 27; Repertory of Claims in Waltham and New Forests, temp. Charles II., 1837 Report, p. 13 a; Volume of Collections as to Forest Law, *Maynard's MSS.* (Lincoln's Inn), xi. and xii.; Various Inquisitions and Rolls as to Northampton and Oxford Forests, 1837 Report, p. 181 b—183; Reading on the Carta de Foresta and other Forest Laws, *Hales' MSS.* (Lincoln's Inn), vol. xl.; Various Surveys as to Forests, 1819 Report, pp. 9, 33-37, 51, 55, 63, 187, and 196.

[1] A very fine series is preserved at Ipswich. See Historical MSS. Commission, 9th Report, p. 226; and *post*, p. 53.

[2] A list of the places on the Inclosure Awards, from 1757-1837, is in the 26th Report, p. 15; and a list of the awards themselves, from 1756-1853, is in the 27th Report, p. 1.

or has a local Archæological Society, you will carefully consult their "proceedings" first. Much time would be saved if a local digest of the articles relating to each county were compiled, as was done for Norfolk in the *Index to Norfolk Topography*, published by the Index Society.

Before writing a Parish History you cannot do better than go through the "Series of Queries to be answered," which were written by Jos. Hunter, and printed by him in the *Journal of the Archæological Institute* of 1847 (Norwich vol. p. 94).[1]

*　　　*　　　*　　　*　　　*

These, as far as regards descriptions of churches, may be amplified thus, remembering that in taking notes you cannot be too full, either in noting angles or dimensions, for it is far easier to run your pen through superfluous particulars when you are fairly copying your history than to make another journey to the church to get further material you thought unnecessary at your first visit :—

Style or styles of architecture, material or materials.

Tower, spire; if tower, whether square, round, or octagonal, &c.

Any chapels or aisles ?

By what windows, nave, aisles, chancel, chapel, and tower, are lit. Any stained glass ? If so, describe it.

What roof ? Of what material ? If old, is it carved or painted ? Has its pitch been altered of recent years ? This you may often trace from the mark of the old roof, if it has.

Any trace of a rood screen, rood loft, or turrets ?

Any mural paintings ? Brasses ?[2] Encaustic tiles ?[3]

How seated or pewed.

---

[1] Vide also the List of Queries submitted to Wiltshire Antiquaries by Hoare.

[2] To enable you to describe them, consult Boutell's *Monumental Brasses*, 1849. Of course there are many detailed accounts of the brasses of various counties, *e.g.*, Cotman's *Brasses of Norfolk*, Dunkin's *Brasses of Cornwall*, Hudson's *Brasses of Northamptonshire*, Belcher's *Kent Brasses*, &c. The Rev. W. F. Creeny's magnificent work on *Continental Brasses* may be consulted with advantage by anyone who suspects a foreign origin for any brass.

[3] See *Examples of Decorative Tile termed Encaustic*, 4to. 100 plates, by J. G. Nichols, 1845, and Shaw's *Specimens of Pavements*, Pickering, 1858, 4to.

What bells?[1] formerly frames for more? What inscriptions? What font? What church plate?[2] Piscina or sedilia? What chest? Any books printed or MS. in it?

Any armour, hatchments, or banners?

Note *all* monuments and heraldry.[3] Charities.[4]

As you are strong be merciful. If you can restrain yourself, *don't* discover that your church is of rather earlier date than St. Martin's at Canterbury, or is founded on the site of a Roman temple. You may be right, but to declare yourself will in all probability destroy your credit as a trustworthy topographer.

CHAPTER III.

𝔇𝔬𝔠𝔲𝔪𝔢𝔫𝔱𝔰 𝔯𝔢𝔩𝔞𝔱𝔦𝔫𝔤 𝔱𝔬 𝔱𝔥𝔢 𝔖𝔲𝔟𝔦𝔫𝔣𝔢𝔲𝔡𝔞𝔱𝔦𝔬𝔫, 𝔖𝔞𝔩𝔢, 𝔞𝔫𝔡 𝔗𝔯𝔞𝔫𝔰𝔣𝔢𝔯 𝔬𝔣 𝔏𝔞𝔫𝔡.

THE most important document of this class, of course, is *Domesday Book*, 1085-6,[5] the nature of which is too well known for me to describe it.

[1] See L'Estrange's *Church Bells of Norfolk;* Raven's, *of Suffolk;* Tyssen's, *of Sussex;* Ellacombe's, *of Devonshire;* Stahlschmidt's, *of Surrey, Kent, &c.*

[2] *The Old Church Plate of the Diocese of Carlisle,* by Ferguson, 8vo. 1882, is a good book on the subject, as are Manning's *Norfolk Church Plate,* Cripps' *Old English Plate,* Fallow and Hope's *York Church Plate.*

[3] See Blore's *Sepulchral Antiquities,* 1826; Boutell's *Church Monuments, &c.* (1854; sells for about 15s.) Farrer's *Church Heraldry of Norfolk* is a model for the collector.

[4] Very excellent accounts of the charities of each parish will be found in one or other of the *Reports of the Commissioners for Enquiring Concerning Charities.*

[5] For a facsimile of part of it see 1819 Report, No. **xxx.**, and of the whole see the recent photo-zincographic edition referred to hereafter.

It comprises all England, except Northumberland and Durham, though it is imperfect for Cumberland, Westmoreland, and Lancashire, which are not described under their proper counties. The counties of Essex, Norfolk, and Suffolk, are far more fully given than any other counties, and are supposed to be practically transcripts of the original rolls,[1] and not merely abstracts of them, as would seem to be the case in the other counties.

The text of the whole was published by the Government in 1783, in 2 vols. fo., to which in 1811 were added another vol. of Indices with able introductions, and a vol. of supplement containing the Exon., Ely, and Winton *Domesdays*, and the *Boldon* (Durham) *Book*, mentioned hereafter. Kelham's *Domesday Book illustrated*, 1788 (now sells for about 18*s.* or more), was the first separate publication on the subject, and is excellent.

In 1833 Sir Henry Ellis published a most valuable work, the *General Introduction to Domesday*, in 2 vols. 8vo., now, unluckily, scarce and dear (£3. 3*s.* or so), which supplemented the folio edition by giving admirable indexes of the tenants in capite, of the persons mentioned as holding lands in the time of the Confessor and before the Survey, and of the undertenants at the Survey, besides a most valuable preface dealing with the general scope of the work, and also going into minute explanatory details. The most excellent handbook to the study of Domesday is, however, the Rev. R. W. Eyton's *Key to Domesday exemplified by an Analysis* (Taylor, 10, Little Queen Street, 1878, now sells for about 10*s. 6d.*) He also wrote *Domesday Studies*, 1880-1, being Analyses of the Somerset and Staffordshire Surveys.

D'Anisy's *Recherches sur le Domesday* (Caen, 1842, sells for 10*s.*) is also a useful work, but was never completed.

Of late years a facsimile edition, called the *Ordnance Survey Edition of Domesday*, has been brought out by the Government in photo-zincography, and is published in counties by the Ordnance Survey Office, Southampton (Agents, Stanfords, Charing Cross.)

---

[1] *Cotton MS.*, Tiberius A. vi. fo. 38, is thought to be a transcript of some others, giving the lists of some of the jurors. Similar transcripts exist for Cambridge, and have been published by Mr. Hamilton.

Some of the counties have had separate translation-extensions published, and some separate treatises have been issued, very full accounts of which are about to be published by the Domesday Society,[1] and by the Royal Historical Society, which has just issued Part 1. of *Domesday Studies.*

Besides *Domesday* proper, there are :—

(*a*) *The Exon Domesday,* in custody of the Dean and Chapter of Westminster, which contains an account of Wilts, Dorset, Somerset, Devon, and Cornwall, and is supposed to be a transcript of original rolls or returns made by Commissioners for Domesday, from which the latter was compiled. It is described in detail in *Cooper on Public Records,* vol. ii. pp. 208-221. For facsimile of part see 1819 Report, No. xxxi.

(*b*) *The Inquisitio Eliensis* (Cotton MSS. Tiberius A. VI.) Supposed to be similar rolls for the possessions of the Monastery of Ely. Refers to the counties of Cambridge, Herts, Essex, Norfolk, Suffolk, and Huntingdon. Described in *Cooper,* i. pp. 222-224. For facsimile of part see 1819 Report, No. xxxii.

(*c*) *The Winton Domesday* (now belonging to the Society of Antiquaries). An inquest taken between 1107 and 1128 for Henry I., who was desirous of ascertaining what Edward the Confessor held in Winchester as of his own demesne. Described in *Cooper,* i. pp. 222-226. For facsimile of part see 1819 Report, No. xxxiii.

(*d*) *The Boldon Book.* Survey of possessions of the Bishopric of Durham, taken 1183. Described in *Cooper,* i. pp. 226-231. For facsimile of part see 1819 Report, No. xxxiv.

The four books above mentioned were printed by Government, with the edition of Domesday published in folio, as above.

Besides these "Domesday Books" there are several others so called, *e.g.*, *The Domesday of St. Paul's* (d. 1222), which was printed by the late Archdeacon Hall in 1858, for the Camd. Society; *The Norwich Domesday; The Ipswich Domesday;* and *The Domesday of Chester,* in the possession of the Dean and Chapter of York. See *Cooper,* ii. pp. 315-347.

Next in date and importance to Domesday came the *Pipe*

---

[1] That is if it ever comes to anything. It seems standing still just now.

Rolls,[1] which are perhaps, all things considered, the most interesting series of records extant.

For each year there is a great brown roll, broad, long, and unwieldy, containing as a general rule as many skins as counties, though sometimes, when the year's matter more than fills both sides of the skin, there is what is called a residuum carried over to some other partly vacant skin. These rolls[2] are mostly in good preservation, and the writing is clear and regular; but the words are abbreviated in a most extraordinary way. The series from Henry II. is almost perfect.

Some of the rolls have been printed: those marked * by the Government, and the others by the Pipe Roll Society.

* 31 Henry I.[3]	8 Henry II.
* 2 Henry II.	9 Henry II.
* 3 Henry II.	10 Henry II.
* 4 Henry II.	11 Henry II.
5 Henry II.	12 Henry II.
6 Henry II.	* 1 Richard I.
7 Henry II.	* 3 John.

The Pipe Roll Society,[1] which is most deserving of support, will continue to print the Rolls of Henry II., and other records

[1] No student of the Records should be without Madox's *History of the Exchequer*, which contains a most elaborate and admirable account of the Pipe Rolls. Nor should he fail to consult Stapleton's work on the *Great Roll of the Norman Exchequer*, published by the Society of Antiquaries in 1840 (1180-1203.) The learned introduction is very useful to the student of the English Pipe Roll, but unluckily it is unindexed; the third volume, which was to have contained the index, never having appeared. Why, it is hard to say, considering how wealthy the Society is, and one is tempted to say with Arthur Orton, "Some folks have money," &c.

[2] There are duplicates of these Rolls, called "Chancellor's Rolls," from 11 Henry II., which were sent to the British Museum, but are now returned

[3] This Roll has been ascribed by Madox to 5 Henry I.; by Prynne, to 18 Henry I.; and by D'Ewes, to 5 Stephen. Others think, as I do, that it is made up of odd membranes of early rolls.

[4] Hon. Sec., J. Greenstreet, Esq., 16, Glenwood Road, Catford, S.E.; Treasurer, W. D. Selby, Esq., Public Record Office; Publishers, Wymans, 72, Great Queen Street. Every student should join, for the subscription is only £1. 1s.; and the extra vol., containing the glossary, is simply invaluable.

to 1200. It published in 1884 a most useful introduction to the study of the Pipe Rolls, with most copious lists of abbreviations and a glossary, as well as a treatise on the usages of the Exchequer.

Oxford students may care to be reminded that Nos. 12-17 of the Dodsworth MSS.,[1] No. 4154-9, Bodleian Library, are transcripts, or rather full extracts, in six folio vols., from early Pipe Rolls.

On the next four pages is given, in the form of a modern balance sheet,[2] an analysis of so much as relates to London and Middlesex of the first of the long series of these Pipe Rolls, as to the date of which there is no dispute (2 Henry II.)

It is hardly necessary to say that the " Pipe " Roll did not get its name from having anything to do with pipes in the modern sense of the word, any more than the Feet of "Fines" had to do with penalties. It used to be called the Great Roll, *the* roll *par excellence*, and its derivation is thought to have been because the sheriffs all sent in their accounts as it were through so many pipes into the common receptacle or Exchequer. This derivation seems rather a far-fetched one, but I have no better to offer.

Taken as a whole the Pipe Rolls are the most interesting of all our national records. To a great extent they are the budgets and balance sheets of the ancient Chancellors of the Exchequer. They comprise yearly accounts of all the taxes collected in the different counties of England, of fines, reliefs, escuages, &c., paid by the tenants in capite, whereby their descents may be easily traced, of sums paid to the king for having justice, or for liberty to commence suits at law.

But to the antiquary, as compared with the genealogist, the interest lies in the entries for disbursements made by the various sheriffs, either on their own responsibilities or by the king's writs. Practically speaking, the king had a banking account open with the sheriff of every county. Whenever the king wanted to make a payment, he gave a close letter on the sheriff of the handiest county, or, as we should say, drew a

[1] *A propros* of the Dodsworth MSS., it is to be noted that an index to the first seven vols. was printed in 1879, but is not for sale.

[2] Another attempt to reproduce a Pipe Roll in a tabular form is on p. 145 of *Thomas' Handbook*, London, 1853.

Gervase and John [Sheriffs of London] render [their] Account

	£.	s.	d.
[To Three Quarters of the Annual Farm]	[407	9	11]

[MEMORANDUM.]

The weavers of London owe two marks
    of gold for their guild.
The bakers of London owe one mark
    of gold for their guild.

	£.	s.	d.
	[407	9	11]
[To the aid of the city	120	0	0

[MEMORANDUM.]

Robert de Ponte owes 20s. for having a plea against Wido fil. Tecii.
William fil. Folcredi owes 40s. for having a plea.

	£.	s.	d.
[To error in casting]	0	6	10
	£120	6	10

DON.

	£.	s.	d.
By the alms lately settled on the Knights of the Temple	0	13	1
By small [fixed payments] granted	10	5	3½
By the [annual] allowance of William fil. Otho	13	13	9
" " William fil. Ailward	8	11	5
By [paid for] char coal for the king's goldsmith	2	5	7¼
By the [annual] allowance of Henry the forester	5	14	0
" " Nathaniel the keeper of the king's houses at Westminster	7	19	8
By oil [bought] for the Queen's lamps	1	2	9½
By [alms given to] the infirm of London	3	0	0
By paid [for] 4 loads of iron for the fortification of the tower	6	15	10½
" 200 ells of flax web to make napkins [as directed] by the King's writ	2	10	0
By spent in carrying the king's treasure to Shoreham	1	6	8
By [paid for] shields and saddles for the King's use to Ernald the shield maker, as directed by the King's writ	10	0	0
By [paid] for orphreys for the King's use as do.	1	0	0
By [paid for] two cushions for do., do.	16	13	4
By [paid for] the presents which the King sent to the Kings of Norway, and for gifts for their ambassadors	37	2	8
By [paid for] making a pavilion for the King and the cost of it, besides the gold ...	96	12	8
By [paid for] the Queen's affairs [as directed] by a writ from her and the justices	3	8	4
By [paid to] Henry the Forester for repairing the gaol	10	0	0
By [paid for] repairs to the King's houses at Westminster [as directed] by the Bishop of Ely	0	5	0
By gold to gild the King's bridles as do.	2	16	0
By delivered out of the Treasury to Roger the doorkeeper [as directed] by the King's writ	1	8	5
By [paid for] repairing the Houses of the Exchequer	3	6	8
By given to Humfry Pincewerre, the King's approver	1	6	3½
" Pollard	1	16	8¼
" Pipelorious	0	15	1
" Robert Crispus the approver	0	16	3
" William fil. Warin the approver	0	6	8
By the allowance of Gerard the forger	1	4	9½
By the expense of the trial of those who forged the King's seal	0	11	2
By paid the expense of the trial of Richard Finch	0	5	5
" to a certain female approver	0	15	2½
" for a house or hut to burn a thief	0	13	4
" for mutilating a maker of false money	0	6	8
" " William Osmund	0	5	0
" " a man beaten in a duel	0	5	0
" for the Queen's corrody [table expenditure]	40	0	0
" for the corrody of the King's son Henry and his sister and her aunt ...	24	0	0
" for wine for their use by [direction of] Ralph de Hastings	7	0	0
" for their corrody by do.	6	6	0
	336	8	1
By balance due to the King	71	1	10
	407	9	11

[By cash paid] into the Treasury		55	10	6
By payments to the King's merchants, through William Cumin		30	0	0
By exemption allowed by the King's writ on the lands of the Chancellor	1 1 8			
" " " the Bishop of Ely	2 0 0			
" " " the King's brother Wm.	3 3 0			
" " " Warin fil' Gerald	3 8 8			
" " " Henry de Essex	1 6 8			
" " " Richard de Lucy	0 13 4			
" " " the Queen's cordwainer	1 0 0			
" " " the Sheriff ...	2 0 0			
		11	16	4
By allowance made for lands [wasted [in the time of King Stephen, and now unable to pay their proportion]		20	0	0
		£120	6	10

Gervase and John render their

£. s. d.

[To amount of the Danegeld ?]   ...   [85 0 6 ?]

£85 0 6

The said Sheriffs in respect of the Fines

To fines receivable during the Shrievalty of Gregory   2 9 10

£2 9 10

The said Sheriffs render their Account of the Gift (aid)

To amount of such gift and fines   ...   58 6 8

£58 6 8

ESEX.

Account of the Danegeld to the King.

				£.	s.	d.	£.	s.	d.
[By cash paid] into the Treasury					...		39	7	6
By exemptions allowed by the King's writ on the lands of Regd. de St. Walery			11	0	0				
,,	,,	,,	the Monks of Bec	2	18	0			
,,	,,	,,	Ernald de Ribemunt	1	10	0			
,,	,,	,,	Martin de Capell	0	4	0			
,,	for the superhidage of the Archbishop	...	...	13	0	0			
,,	on the lands of the Barons of Wallingford	...	...	1	8	0			
,,	,,	the Bishop of Ely	... ...	0	2	0			
,,	,,	Jordan de Samford	... ...	0	10	0			
,,	,,	Henry de Essex ...	... ...	0	4	0			
,,	,,	Ralph de Hastings	... ...	0	7	0			
,,	,,	the Sheriffs	... ...	1	5	0			
,,	,,	the infirm	... ...	0	2	0			
,,	,,	the Canons of the Holy Trinity	.. ...	0	3	0			
							35	13	0
[By allowance made for lands] wasted [in the time of King Stephen *ut ante*]				...			10	0	0
							£85	0	6

...yable for Pleas and Murders.

	£	s	d
By balance due to the King	2	9	10
	£2	9	10

... the County, and of the Fines for Pleas and Murders.

				£	s	d	£	s	d
[By cash paid] into the Treasury					...		6	13	4
By exemptions allowed by the King's writs to the Archbishop ...			...	8	6	8			
,,	,,	,,	the Monks of Bec	...	2	8	0			
,,	,,	,,	Reginald de St. Walery	...	10	0	0			
,,	,,	,,	Robert fil. Sawini	...	0	3	6			
,,	,,	,,	the infirm of Westminster		0	3	7			
,,	,,	,,	the Barons of Wallingford		3	17	8			
,,	,,	,,	Henry de Essex ...	...	0	3	6			
,,	,,	,,	the Bishop of Ely	...	0	1	9			
,,	,,	,,	Martin de Capell	...	0	3	6			
,,	,,	,,	William fil. Otho	...	0	10	6			
,,	,,	,,	the Prebends of Westminster	0	3	6			
,,	,,	,,	the Sheriffs	... ...	1	0	0			
,,	,,	,,	Gregory	... ...	0	13	4			
							27	15	6
By paid to the King's sons before the Queen crossed the sea	...		...				4	0	4
By the payments to the Knights of Wallingford and Hereford	...		...				16	13	4
,,	,,	Templars	... ...		...		0	13	4
							55	15	10
[By balance due to the King] ...							2	10	10
							£58	6	8

cheque on him. He was not above overdrawing his account, by the way, for we not infrequently find the payments made greater than the creditor side of the account; and then the sheriff "habet de superplus (surplus)" for so much money, and this was seemingly paid him by the sheriff of the following year, who enters it as a first charge on the new year's income.

The sheriff also deducted for the lands which originally contributed to the "farm" of the county, but which the king had afterwards given away. Then he accounted for all payments made by him for repairing the king's houses or castles, for executing criminals and moving them about from place to place, for entertaining ambassadors, for rewarding those who brought up the heads of outlaws, for purchase and carriage of wine and other necessaries for the king's table, for supplying venison for the king's table, and so on. The entries which occur in this part of the roll are of the highest antiquarian interest, and if it is thought too great a task to publish the whole of the roll for the county, it would be very interesting to go through the earlier rolls and extract these items, for they cover a very interesting period before the date of the earliest of the Close Rolls, which begin in the 6th John.

I cannot say possibly as to Middlesex, but as to several other counties, I know as a fact, that the interesting information just referred to as contained in these rolls has never been printed, or, as it were, posted up into the county histories.

Taking the printed sheet as a sort of text for my remarks, the first thing that strikes us is the "alms" of 13s. 4d. given to the Templars. They were then at Holborn, not moving to Fleet Street till 1184. Then we come to the annual allowances to William Fitz Otho and William Fitz Ailward. The former is calculated at 9d. a day, but I cannot make out at what rate the latter is reckoned. Henry the Forester received at the rate of 9s. 6d. a month, while the pension of Nathaniel, the Keeper of the King's Houses at Westminster, would seem to be 6½d. a day.

The oil bought for the Queen's lamps was probably to supply votive lights burning before altars, and the infirm of London were probably the occupants of some favoured hospital.

Similarly "the infirm of Westminster," who occur later on, may have been the inmates of the Hospital of St. James, on the site of which St. James' Palace now stands. The next entry, which abbreviated runs "Et pro quatuor cumb' fr' ad munitionem Turris," I cannot translate to my liking. "Quatuor cumbæ frumenti" would be the natural reading, but the price (£6. 15s. 10½d.) seems to put this out of the question. Fr. is probably for iron.

The amount paid for moving the king's treasure to Shoreham would seem to prove that Old Shoreham in Sussex was probably meant, though, as there was formerly a castle at the Shoreham near Sevenoaks in Kent, the king may have been staying there.

Ten pounds for shields and saddles, and no less than £96.12s.8d. for a pavilion, apparently gilt, are large sums, considering the then value of money.

Humfrey Pincewerre, who is called the *King's* approver, was probably so called because he had become what was afterwards called "king's evidence," though he may have been one charged with letting out the King's manors.

"Pollardus" may have been a personal name, and so may have been "Pipelorius," but considering the company in which they occur, it is curious to note that "Pollardus" is an obsolete word for debased money, and that "Pipolo" is given by Ducange as having the meaning of "convictio."

The entry "pro una domo ad comburendum unum latronem" is a very curious one, and it is hard to say in what sense we are to read "domus" here. One can hardly suppose that our ancestors, like the roast pig burners celebrated by Charles Lamb, had not yet discovered that a man could be burnt to death without sacrificing a house, and all I can suggest is that something in the nature of a hut or hovel (let us hope of green wood, so that he was speedily suffocated), was built round the stake, so as to hide the execution.

The only thing at all bearing on the subject I have been able to find is in the *Laws of the Miners of Mendip*, published in 1687, quoted in the 1st series of *Notes and Queries*, vol. ii. p. 498, where it is laid down that anyone stealing ore of the value of

13½*d.* was to be put with his tools into his house, and everything set on fire about him.

No punishment could indeed be more cruel than those inflicted under our early kings. The next entries refer unmistakeably to the barbarous mutilation of three prisoners, one a forger of the King's money, and another the unhappy defeated combatant in a wager of battle, who was, of course, doomed to pass the rest of his life in a cloister.

It is noteworthy, by the way, that this roll twice refers to Henry de Essex, the King's Standard Bearer, as being high in favour and receiving grants and remissions from the King. Only a very few years after he too was accused of cowardice, defeated in a duel, and relegated to Reading Abbey.

The entries, both in the "Aid" and in the "Danegeld," of allowances made for "Wasta," are very noteworthy, and mark how severe must have been the sufferings inflicted by the recent wars in the time of Stephen. £20 are remitted out of £120 for the Aid, and £10 out of the £85 odd of the Danegeld. The same thing is shown by the shortness of the roll and the paucity of the entries, compared with the earlier roll usually ascribed to 31 Henry I.

In the Danegeld account the allowance to the Archbishop of "superhidage" proves that the Danegeld was originally rated at so much a hide, and that by some means or another, either by mistake or from the Archbishop having sold some of his lands, he had been debited with too much, and that this overcharge was allowed him by way of deduction.

The Barons or Knights of Warengeford I take to have been the "Barons" of Wallingford, but speak under all correction.

A few words to conclude with as to the accountant's work on these rolls, may not be out of place. The roll, so far as relates to London, is not a favourable specimen of early city bookkeeping. In the account of the farm of London the debtor amount is not set out, though from the creditor items coming to £407. 9*s.* 11*d.* for the three-quarters of the year, it would seem that yearly farm must have been about £543.

In the account of the Aid there is an error of 1*s.* in the casting of the exemptions, which should be £14. 16s. 4d., and not £14. 15s. 4d. as on the roll, and the whole account does not balance by 6s. 10d. It is possible, however, that this apparent surplus may have been for the dealbation[1] of some part of the payments into the Treasury; but I do not think it is likely. as the dealbation is apparently always *specially* mentioned, and the only mention of blanched silver is in the statement of the farm of the city where the balance due to the King is said to be £71. 1s. 10d. "blanch," whereas my casting makes it £71. 15s. 3½d.

The account of the Aid of the county both casts up and balances correctly.

Connected, to a certain extent, with the Pipe Rolls are the class of documents known as the *Miscellanea of the Exchequer*, of which there are seventeen vols. of MS. Calendar in the Round Search-room at the Record Office. They include all manner of documents, such as inquisitions, inventories, &c. For a report on them see 1st Report, p. 123 et seq.; and on the *Miscellaneous Records of the Queen's Remembrancer of the Exchequer* see 40th Report, p. 467. Cole published in 1844 a volume styled *Documents illustrative of English History in the Thirteenth and Fourteenth Centuries, obtained from Records in the Exchequer*.

From time to time, on great occasions or epochs when the collections of important *Aids* were about to be made, elaborate books (which belong to the Miscellanea just referred to) containing statements of all who held *in capite* of the King, and to a certain extent of subinfeudations, were compiled by the clerks of the Exchequer for the use of the collectors, partly from returns made by the tenants and partly from the information given by the earlier Pipe Rolls.

These books bear some resemblance to Domesday, in so far as they are divided into counties, but give no descriptions of the land, which is only specified by the knights' fees.[2]

---

[1] The reckoning at assay value.

[2] There are collections for various counties, *e.g.*, Essex. See 1837 Report, p. 195 a.

The earliest of them is the *Liber Niger Seaeearii*, compiled temp. Henry III. It is made up (*i.a.*) of the returns of the tenants *in capite* in 1166, who certify how many knights' fees they hold and the names of those who hold or held them. This has been printed twice, once by Hearne in 1728, and again in London in 1771. A detailed analysis of its other contents will be found on pp. 168-9 of Thomas' *Handbook*, and descriptions of the MS. are on pp. 39-46 of Sim's *Manual*, and in the 1837 Rep. p. 167.

Liber Rubeus Scaeearii, compiled about the same time as the last, contains another copy of the 1166 returns, and much other matter from the Pipe Rolls and other sources. It has never been printed,[1] but is fully described in Sims' *Manual*, p. 41; Thomas' *Handbook*, p. 255; and in the Record Report of 1837, pp. 166-177. A separate account of it was printed by Hunter in the 1837 Report, and in his "three Indexes." It contains the only known fragment of the Pipe Roll of 1 Henry II., and copies of the important Inquisition returned into the Exchequer in 13 John.

The Testa de Nevill or *Liber Feodorum* is a similar book, which contains an account of those who held of the king in capite, and has been printed in one fo. vol. by the Record Commission in 1807; but the words on the title page, temp. Hen. III.-Edw. I., are most misleading, for some of the returns are as early as Richard I., and many are of John. Roughly speaking, they are all prior to 1250. Cooper, i. p. 259, considers it was compiled near the close of Edward II. or the beginning of Edward III., partly from inquests taken on the presentments of jurors of Hundreds before the Justices Itinerant, and partly from inquisitions upon writs to the Sheriffs for collecting scutages, aids, &c. This document is not "of record," *i.e.*, not evidence (Rot. Parl. ii. pp. 70, 71). For a facsimile of part of it see 1819 Report, No. xxv.

The MS. preserved in the Record Office, and known as *The Book of Aids*, is dated 20 Edward III. (1327), and is a splendidly written MS. of 318 pp., giving an account of the knights' fees for nearly every county.

---

[1] I understand Mr. W. D. Selby is likely to print most of it for the Rolls Series. The Dialogus de Scaccario has been printed by Stubbs in his *Select Charters*, by Madox in his *Exchequer*, and by Mr. Wren the "coach."

This has never been printed, except for Norfolk, which I published in vol. i. of the *Norfolk Antiquarian Miscellany*, and for Kent in the *Archæologia Cantiana*.

There may also be found much about knights' fees in the Crown Plea Rolls, in which there are numerous presentments like these :—[1]

"That William, son of Isaac de Felmingham, holds lands worth 40s. a year in Bekham of the king, but they do not know by what warrant, therefore let it be enquired into."

"That Richard de Playz holds a knight's fee, and is of full age, but is not yet a knight ; therefore," &c.

The *Hundred Rolls*, taken under a Commission of 2 Edward I. (1274), which is said to have been issued to enquire about abuses which had grown up during the reign of Henry III., also contain much about knights' fees, though primarily the presentments are of the claims and unjust claims of privileges such as free warren, frankpledge, and assizes of bread and ale, and so on. They were printed by the Record Commission in folio (2 vols.), and can now be bought for about £4. A fac-simile of entries on them will be found in 1819 Report, Nos. lv. and lvi.

The *Placita de Quo Warranto*, taken a few years afterwards, which have also been printed in one vol. fo.. may be bought now for about £2. 10s. A facsimile is in the 1819 Report, No. lvii.

Kirby's Quest, taken 35 Edward I., was an enquiry as to the tenures of tenants in capite of certain northern counties, and the Yorkshire portion has been well printed by the Surtees Society in 1867.

The next class of Records which relate, more or less, to knights' fees and subinfeudation are the *Oblata* or *Fine Rolls*, which are records of the Chancery upon which writs of *diem clausit extremum*, of seizin on heirs doing their fealty or homage for the lands of a deceased,[2] custody of heirs, liveries of seizin,

---

[1] Crown Plea Roll, Norfolk ; 34 Henry III. $^{m\,41}_{1}$

[2] The Inquisitions Post Mortem themselves are referred to hereafter.

of dower, license to remarry, to leave England, to have letters of protection, and so on. The Oblata begin 1 John, and the Fines in 6 John.

Selections of these Fine Rolls have been published by the Government under the following titles :—

Rotuli de Oblatis et Finibus in Turri Londiniensi asservatis tempore Regis Johannis, 1 vol. 8vo., 1835 ;

Excerpta è Rotulis Finium in Turri Londiniensi asservatis temp. Henry III., 2 vols. 8vo. 1835-6.

There is a MS. continuation of the extracts for the reigns of Edward I. and part of Edward II. in the Round Room of the Record Office, with Indexes Nominum only, and not Locorum. A specimen of a short tabular calendar of a Fine Roll is at p. 74 of the 1837 Report.

If the tenant in capite wished to give away land to a monastery, or if anyone begged for the grant of a fair or a market, the question whether the gift or grant would be injurious to the Crown was tried on an *Inquisition ad quod damnum*, a calendar of which from 1 Edward II. to 38 Henry VI. has been printed, forming Part 2 of the *Calendarium Rot. Chartar.*, the whole vol. now selling for about 15s.

CHAPTER IV.

The Sale and Transfer of Land inter Vivos.

THE grant or charter[1] formed the earliest and simplest way of transferring land, and of early charters immense numbers are preserved in various places.

Many Saxon charters were collected by Kemble in his *Codex*

[1] For a specimen form of charter see p. 144.

Diplomaticus; by Thorpe (under the title *Diplomatarium Anglicum Ævi Saxonici*, 1865); and a work, "*Cartularium Saxonicum*," is now being published by the voluminous Mr. Walter [de Gray] Birch, which also treats on the same subject, but is of no great value. Other Saxon charters have been facsimiled by the photo-zincograph process at the Ordnance Survey Office. See 45th Report, p. 381, and 46th Report, pp. 127 and 403. The Pipe Roll Society are about to publish a volume of early charters.

Thousands of charters of various ages are in the British Museum among the Additional and other MSS., and are partly calendared in the printed indexes.

A general MS. Index to Charters is now in existence, and can be consulted in the Record Office Search Room, but in the Great Room of the MSS. Department are much larger classified catalogues of MSS., including even individual letters, as well as charters.

The charters and rolls, preserved in the Bodleian Library at Oxford, were calendared by the late Mr. W. H. Turner, though not too well. The volume was published in 1878, and has 150 pp. of index in double columns and small type.

At Lambeth there are thirteen fo. vols. of charters, &c.

At the Record Office there is a series of forty-one rolls, called the *Cartæ Antiquæ*,[1] consisting of 12th and 13th century transcripts of royal and other charters from Ethelbert to Henry III. Of these a calendar was published by Sir Joseph Ayloffe (4to. 1774), the title being *A Calendar of Ancient Charters and Scotch and Welsh Rolls in the Tower of London*, which now sells for about 7s.

The enrolments on the Charter Rolls[2] (1199-1483) practically form a continuation of this series. Those for King John were published in full by the Record Commission in fo. in 1837, *Rotuli Chartarum in Turr. London Asserv*. A calendar of the whole series in 1803, under the title of *Calendarum Rotulorum Chartarum et Inquis. ad quod damnum*, &c., now sells for about 15s.

[1] A modern transcript of the *Cartæ Antiquæ* is among the transcripts kept in the Round Room.

[2] There was a Norman series of Charter Rolls, which will be found referred to in the *Rot. Normann. in Turr. Lond*, see *Rec. Report*, 1835, *and* 41st Report, p. 671, &c.

All these are Royal Charters, no private deeds being among them.

For a modern calendar of early Royal Charters, see 29th Report, pp. 7-48, and 30th Report, pp. 197-211.

There is also another collection, also called *Cartæ Antiquæ*, formerly in the Augmentation Office, which are the deeds belonging to the dissolved monasteries. See list in 20th Report, p. 79 ; but the most valuable collection of original deeds will be found among the Chapter-house Records. A calendar of them is now in progress. The Chancery and Augmentation Office Deeds are also being calendared. See 48th Report, p. 8.

Other deeds were also enrolled for safe custody on many of the regular series of Rolls, even on the Pipe Roll.

There are many on the De Banco Rolls (Common Pleas), and of these there is an " Index " from Michaelmas, 20 Henry VII., but only under counties, without any lexicographical or even alphabetical arrangement. Still it is better than nothing. The first part of this index or rather calendar (20 Henry VII. to 31 Henry VIII.) is contained on pp. 124-157 and pp. 160-214 of the first vol. of the *Recovery Indexes* mentioned hereafter; vol. i. of the *Deed Index*, beginning 1555-1629, and containing 352 pp. There is also a " Deed Index," 5 vols., 1555-1836, referring to deeds on the Common and Recovery Rolls. A printed inventory of them will be found in the 3rd Report, Appendix ii. p. 130, and see 1837 Report, p. 142 b.

There are also some enrolled on the Queen's Bench Rolls, and a great many on the dorses of the Close Rolls.

Of the more recent of these there are 68 vols. of calendars in the Long Room, beginning 1574 and ending 1881, which from 1698 give the counties. This is a most valuable series.

Nearly every city of any importance had its Register, into which were copied all the deeds either from or to the corporation ; while monastic chartularies are even more common, and will be dealt with later on.

A form of the ordinary charter and its endorsed livery of seizin will be found in Appendix I., p. 144.

By Acts of Parliament passed in the reigns of Queen Anne

and George II. Registries of Deeds were constituted for the counties of Middlesex and Yorkshire.

The *Middlesex* Registry is in Great James Street, Bedford Row, and the searcher will find in Mr. Stahlschmidt, the Deputy-Registrar (who is himself a well-known and able antiquary), a most efficient and courteous guide. The earlier indexes are alphabetical, and of surnames only; but of late years there is an admirable lexicographical index. The fee for searching the old index is 1*s.*, and an extra shilling for looking at every three documents. Nothing may be copied.

The *Yorkshire* Registry dates from about the same period, and has branches at Wakefield, Northallerton, and Beverley.

So much for the Charters and Deeds. We now come to a series of documents which cover an unbroken period of over six centuries, and have no parallel in the records of any other country. I mean of course the *Feet of Fines*, which begin in the reign of Richard I., and which were practically deeds transferring land, though nominally the "finis" or end of a fictitious[1] suit.

Fines,[2] which operated nominally as an amicable arrangement, putting an end (finis) to hostile suit in the King's Court, early became a popular method of conveyance, not only from their efficacy but from the safety ensured to a purchaser, by the fact of a duplicate of each fine being preserved of record in the custody of the court. These duplicates being preserved and arranged in counties, might have served as the germ of a general Law Registry; and indeed, before the Act for Abolition of Fines and Recoveries, the solicitor of a purchaser of landed property used to search in the vendor's name, just as he now does in the Middlesex or York Registry.

[1] Occasionally, however, a real or litigious suit was ended by a Fine or Final Concord.

[2] For a very learned account of the history and method of levying Fines and Recoveries, see Cruise's *Essay on Fines and Recoveries*, third edition, London, 1794. But until the Fines are printed or calendared no real account of them can be compiled. In some notes to my first part of *A Calendar of Norfolk Fines*, for example, I pointed out that it is quite a mistake to suppose Fines related exclusively to real property, and I cited many instances in which they referred to remission of debt, remission of services, freedom of villeins, and admission to benefit of priories, and so on.

On their value to the topographer, and to a less degree to the genealogist, I need not dilate, though I may remind my readers that the earlier fines give minute accounts of the transfer of advowsons, manors, &c., at a date when deeds are excessively rare.

Occasionally we find in them entries interesting for other reasons. Thus, Humphry de Herlham in 8 Ric. I. sold, for a hundred shillings, land to Ralph de Herlham (possibly his brother) for his equipment to Jerusalem, probably in the fifth Crusade then being instituted by Pope Innocent III.; and the objection naturally felt by large landowners to their sub-tenants giving land into mortmain is well instanced by Eva, daughter of Hawis de Morlai, granting lands to Ralph de Torcy and his assigns—"præterquam viris religiosis." The nominal considerations for the grants are sometimes noteworthy—as, a pair of white gloves, a pair of spurs, a pound of cummin, and a pound of pepper, or sixpence.

The early Feet of Fines are written in a very small writing, on pieces of parchment about as long as but narrower than one's hand, indented or notched along the top, and are, as a rule, in good condition; but the following transcript (printed as closely as country type will do it) of one of them will give a better idea of their form and contractions than any description however ample.

H -. final ꝯcord fča In Cuꝛ Dñi Reg̃ ap̃ Norwič Anno Regni ꝶ Rič vijᵒ Die Sabb in festo Sče Marg̃ Corā Witt de Glāvill ⁊ Osb̃ fił Il'uič ⁊ Sim̃ de Patesh Justič Dñi Reg̃ ⁊ alijs fidelib; Dñi Reg̃ ꝶč ibid p̃sentib; Inⱦ Rob de Hicford peⱦ ⁊ Adā de Nereford teñ de . viij . acꝛ č ptiñ ī Spham . uñ ass̃a de morte añcessoris sum̃oñ fuit Iuⱦ cos i cuꝛ p̃fata: Sciłt qđ iđ Adā cōcessit ciđ Rob ⁊ hedib; suis totā pđčam ꞇrā cū ptiñ tenenđ de se ⁊ hedib; suis ꞏ Reddendo Iñ añuati sex denaꝛ ad fesⱦ Sči Mich . ⁊ ad xx . soł de Scutag̃ : vi đ ⁊ ad plus ꞏ plus . ⁊ ad miñ ꞏ miñ p ŏi ꝯsuetudine : ⁊ p hᵃ cōcess̃one iđ Rob deđ dčo Ade tria Bizantia   Norf

The above extended reads thus—

Hæc est finalis concordia facta in curia Domini Regis apud

Norwicum anno regni Regis Ricardi septimo die Sabbati in
festo Sancte Margarete Coram Willelmo de Glanvill et Osberto
filio Hervici et Simone de Pateshull Justiciariis Domini Regis
et aliis fidelibus Domini Regis &c. ibidem presentibus Inter
Robertum de Hickford petentem et Adam de Nereford
tenentem de 8 acris cum pertinentibus in Sparham. Unde
Assisa de morte antecessoris summonita fuit inter eos in
curia prefata; scilicet quod idem Adam concessit eidem
Roberti et heredibus suis totam predictam terram cum pertinentibus Tenendam de se et heredibus suis Reddendo inde
annuatim sex denarios ad festum Sancti Michaelis et ad 20
solidos de scutagio sex denarios et ad plus plus et ad minus
minus pro omni consuetudine. Et pro hac concessione idem
Robertus dedit dicto Ade tria bizantia. Norfolchia.

And translated,[1] thus—

This is the final agreement made in the Court of (our) Lord
the King at Norwich, on the Saturday on the feast of
St. Margaret, in the 7th year of King Richard, before William
de Glanvill, Osbert Fitz Hervey, and Simon de Pateshill,
Justices of (our) Lord the King, and other faithful (servants)
of (our) Lord the King, &c., there present, between Robert
de Hickford (the) demandant, and Adam de Nereford (the)
tenant of eight acres with (their) appurtenances in Sparham,

---

[1] For an extended form of fine see Appendix, p. 142. The modern form
generally ran thus—

This is the final agreement made in the Court of our Sovereign Lord the
King at Westminster in the . . . days of . . . in the . . . year of, &c., before
A B, &c., Justices of our Lord the King and others then and there present ;
between C D, demandant, and E F, tenant, whereupon a plea of covenant
was summoned before them in the said Court, that is to say, that the said
E F has acknowledged the aforesaid tenements to be of the right of him
the said C D, as those which the said C D hath of the gift of the said
E F; and those he has remised and quit claimed from him the said E F ;
and moreover the said E F hath granted for him and his heirs, that they
will warrant to the said C D the aforesaid tenements against him, the said
E F, and his heirs for ever. And for this acknowledgment, remise, quit
claim, warranties, fine, and agreement, the said C D hath given to the said
E F £ . . . sterling.

concerning which an assize of mort d'auncestor was summoned between them in the aforesaid court.

Namely, that the said Adam granted to the said Robert and his heirs all the aforesaid land with (its) appurtenances. To hold of him and his heirs, paying therefore yearly sixpence, and for every twenty shillings of scutage sixpence, and for more more, and for less less, for all service.

And the said Robert for this grant gives to the said Adam three bezants.—Norfolk.

The volumes published by the Record Commission in 1835-44, under the title of *Fines sive Pedes Finium*, relate to the reign of Rich. I. only, reach only from Beds to Dorset; nor from their commencement until 1 Henry VIII. are there any calendars, let alone indexes to them.[1] The "Indexes to the Fines,"[2] which are in the Round Room, consist of forty-five volumes of calendar, which is arranged in terms and counties, the entries running as on the Entry Book (see next page), except that the number and the attorney's name is omitted.

The Calendar goes on to Michaelmas, 32 George II. (vol. 45), where it stops as far as the Round Room is concerned; whether the rest of it is in existence I have been unable to ascertain. Anyhow, there is no more on the shelves, and vols. xlvi. et seq., which serve as a continuation, are really part of another series (of which earlier vols. exist). These later vols. have been improperly bound and lettered as though part of the "Index" series, which is very confusing, and they only give the surnames of the plaintiffs and defendants, and the name of the plaintiffs attending.

The Books of Entries of the Fines, however (most of which

[1] There certainly are a few (eleven) volumes, purporting to be indices to fines, which have been noted by private collectors, chiefly Le Neve, but these refer to so infinitesimally small a portion of the fines that they are almost useless. The most useful of these are vols. xii. b (formerly lxii., xii., and xiii.,) containing references to fines for "divers counties" (*i.e.*, for fines which refer to property in more than one county) from John to Henry VI. by Le Neve, which have indexes. Some counties, it is true, have separate indices to some of their fines. See analysis further on.

[2] Which are bound up in thin bundles of parts of twenty-five.

are in the Round Room[1]), and of which there are fifty-four vols., from 1611-1829, seem to fill up the gap,[2] giving the information thus—

297. Henry Bell esq Plt & William Townley and Mary his wife & others Defts in Wrettone—Millett.

But the chief use of them is that the late volumes of this series give references to the actual number borne by the Fine.

Lastly, the *King's Silver Books*[3] (which exist from the reign of Henry VIII., but owing to damage done them by fire are inconsultable until that of George I.) give yet more information about the property and parties, *e.g.*, the above referred-to entry is thus extended.

20/. *Norfolk* Henry Bell Esq[re] Plt W[m] Townley & Rob[t]   13/4
8 St Hil.   Thorpe & Ann Mary his wife defts of 2
mess 1 barn 1 sta 2 cur 2 gar 1 or 20 La
20 mea 20 pas & com of pass for all mann
of catt with the app[ts] in Wretton & Stoke
ffery Bfore Harvey Goodwin & Jno
Houchen gents by com 29 Decr 38 K Geo 3

Other ways of getting references to fines there are, as by searching the "Covenant Book" and the "Concords of Fines;" but as they are generally more troublesome to find, and give less satisfactory information when found, I think the three series above detailed will suffice.

It may be said here that Durham, being a County Palatine, had its fines separately, which, until recently, were kept up in the North. They are in a very bad state, and have no calendar whatever.

What has been done to calendar these invaluable records, the backbone of every pedigree and county history, is shown

[1] From Hilary, 3 George III.

[2] It may be as well to tell the novice one or two "wrinkles" about searching these books. (*a*) If a fine relates to more than one county, it will be found at the beginning of the term. (*b*) If the property is in a city, it will be found under the separate head of "Civitas," between Cantebr. and Cornub. (*c*) If in a town, under "villa," between Suth. and Warr'. (*d*) If in London, under a separate entry—not under "Middlesex."

[3] The King's Silver (or the Post Fine) was the fine paid to the King for liberty to compromise the imaginary suit.

in the table below, where I may boastfully point out, that the only county—and that, far the heaviest in entries—which has been calendared and indexed[1] throughout is that of Norfolk, which was done, in great part, by the indefatigable Le Neve, and again recently by the present writer.

Beds	
Berks	
Bucks[2]	
Cambridge	Full transcripts of the Fines for Richard I.
Cornwall	and John were printed by Government in
Cumberland	2 vols. 8vo.
Derby	
Devon	
Dorset	
Durham and the rest of England	Full transcripts, unprinted, for these two reigns, are still in the Record Office in MS.

Cambridge.—"Calendar, from Richard I. to Richard III.," by Walter Rye, now preparing for the press for the Cambridge Antiquarian Society, and will soon be published with ample indexes.

Derby.—"A Calendar of the Fines to 42 Henry III." have been published in vols. vii., viii., and ix. of the *Journal of the Derbyshire Archæological and Natural History Society* (Hon. Sec., Dr. Cox, Barton le Street, Malton, Yorks.); the work is proceeding.

Essex.—"MS. Calendar, Richard I. to Richard III." (no index), Record Office, Round Room.

Gloucester.—"Index of Fines for George I.," printed by Sir Thomas Phillipps, Cheltenham, 1865. He also printed a few pp. from 1 John, and from 1649-1714.

[1] The reader must not be misled by the statements contained in the reports of progress on pp. 21-33 of the Report, which speak of "indexes" being compiled. They are only part of a series of "calendars" of later Fines now in the Long Room.

[2] Bucks. There are also two vols. of Calendar, Richard I. to Henry VI. (vols. xvi. and xvii.) with indexes of names and places.

Gloucester.—"MS. Calendar, John to Richard III." (index locorum only), Record Office, Round Room.

Hants.—"MS. Calendar, Richard I. to Edward I.," Record Office, Round Room, vol. xxv.

Herts.—"MS. Calendar, Richard I. to Richard III." (index locorum only), Record Office, Round Room.

Kent.—"MS. Calendar of Fines, from 2 Henry III. to 35 Edward I."; "Lansdowne MSS." 267, 268, British Museum.

"Kent Fines, Edward II." *Arch. Cantiana*, vol. xi., by J. Greenstreet.

Norfolk.—(Le Neve's MS. Calendars, some of which are in the Record Office, are not as easily searched as—

"Pedes Finium relating to the County of Norfolk, from 3 Richard I. to the end of John," by Walter Rye, Norwich, 1881, pp. xix.-154, with ample indexes of names and places (a detailed calendar).

"A Short Calendar of the Feet of Fines for Norfolk in the reigns of Richard I., John, Henry III., and Edward I.," by Walter Rye, Norwich, 1885, pp. 218, part 1, with indexes nominum and locorum.

"A Short Calendar of the Feet of Fines for Norfolk in the reigns of Edward II. to Richard III.," by Walter Rye, Norwich, 1885, pp. 219-502, part 2, with indexes nominum and locorum.

"MS. Calendars," by the same compiler, in his own library, indexed as to surnames to 1760.

Southampton.—"MS. Calendar, Richard I. to Edward I." See Hants.

Wilts.—From 1 Edward III. to Richard III. (partially ?), extracted from Lansdowne MSS. 306-7-8.

"Index, from 1 George I. to 2 George II." Printed by Sir Thomas Phillipps, Middle Hill Press, 1 vol. fo. (sells for £1); also fragments from 1 Edward III. to Richard III.

Worcester.—" Index of Fines, from 1 Edward III. to Henry
VI." Printed by Sir Thomas Phillipps.
" Fines for Charles I." Id.
York.—" Fines temp. John." Dodsworth's MSS. lxxiii.
,, 4-51 Edward III., id.
,, Henry VI., id. cvi.
,, Henry III.—Henry V. "MS. Calendar of
Fines," with index locorum. Record Office,
Round Room, vols. xx., xxi.
Id., Tudor period. Now publishing by the
Yorkshire Archæological and Topographical
Association, vol. ii. (Hon. Sec., S. J. Chadwick,
Esq., F.S.A., Church Street, Dewsbury.)

If my reader wishes to search for any county which is not
lucky enough to possess a special calendar, he will find
Lansdowne MSS. 306, 307, and 308, at the British Museum,
a great help, for the entries are arranged in counties in a
tabular form, and the great labour of searching the file of fines,
which have to be specially written for in bundles of twenty-five,
will be saved. When he finds the fine he wants, he can
note the term and year in which it was levied, and then go
to the Record Office and write for the original file. Why the
Record Office have not long ago had a transcript of these MSS.
made and placed on the shelves of the Round Room, is one
of those mysteries which will probably never be solved. The
three volumes are in all probability the original entry books
which have gone astray from the proper office, and have a
special value of their own, inasmuch as they often record
fines now lost from the files.

Nearly following the Lansdowne MSS., just referred to,
we have the fine Calendars of the Fines, by Le Neve, for
Rich. III. and Henry VII. (vol. xxii. in the Round Room, which
bears a modern and incorrect label, "Divers Counties," and
which is indexed in vol. xxvi.), for Henry VIII. (id., vol. xxiii.
and xxiv., not arranged in counties however), and Edward VI.,
and Philip and Mary (id., vol. xiv., having dissected indexes of
places under counties, and an alphabetical list of surnames.)

For the reigns of Richard I. and John there are, I believe,

full extended transcripts of the fines of all counties, which were prepared for publication in the series which came to an untimely end with Dorset.

If these were indexed and placed on the Search Room shelves, they would be a great boon to those who cannot read the old hand. It seems to me that if it were generally known that plain copies of the fines for these two early reigns were in existence, and could be transcribed for the press by unskilled labour, many county societies would be ready to print those relating to their own county.

Analogous to the Fines, and like them the offspring of a legal fiction, were the *Recoveries*,[1] which were judgments obtained in fictitious suits brought against tenants of the freehold, in consequence of the default made by the person who "vouched to warranty" in the suit.

The farce was gravely enacted thus:—The man who wanted to sell or charge the land allowed the intending purchaser to bring an action against him for it. Instead of defending this action by showing his title, he called upon some one (usually the crier of the court) who was supposed to have sold the land to him at some previous time, and to have then warranted him a good title. The vouchee appeared, and admitted the soft impeachment, and took on himself the defence of the title. Then the plaintiff "craved leave to imparle" with the vouchee, and walked out of court with him. Soon the plaintiff returned; but the vouchee, presumably overcome by the cogency of the plaintiff's arguments, and seeing the hopelessness of defending the title, did not. The Court then solemnly called him thrice to come in and defend, but he never did, so the formal judgment was in favour of the plaintiff: the defendant, at the same time, recovering against the faithless vouchee "lands of equal value" for his breach of warranty, a judgment which did not materially affect the peace of mind of the crier.

Such was the clever scheme invented by the monks to evade the statute of mortmain, for they did not "purchase" the land or have it given them, but only "recovered" what they said

[1] See Cruise on *Fines and Recoveries*, cited *ante*.

was theirs as of old title. An act passed 13 Edward I. crippled this; but afterwards they were brought into use again to evade the Statute de Donis, which had forbidden alienation by tenants in tail, and in 12 Edward IV. it was established that a recovery suffered by a tenant in tail effectually barred his issue and the remainder men. They then, of course, became very frequent.

On the recovery, seizin had to be given by the Sheriff, and the award of the writ of execution was entered on the recovery. The recoveries should have been entered on record, but many were not, especially between the reigns of Anne and George II., which default occasioned the Act 14 George II., cap. 20.

The Remembrance Rolls of the Common Pleas begin 36 Henry VIII.

The inventory of the Recovery Indexes is to be found at p. 127 of 3rd Report, Appendix II. The "index" to the Recoveries begins Michaelmas, 22 Henry VII.; but until 4 Anne, 1705, it is not arranged in counties, and is simply a list. It is in the Long Room.

CHAPTER V.

𝕷𝖊𝖌𝖆𝖑 𝕻𝖗𝖔𝖈𝖊𝖊𝖉𝖎𝖓𝖌𝖘 𝖗𝖊𝖑𝖆𝖙𝖎𝖓𝖌 𝖙𝖔 (1) 𝕷𝖆𝖓𝖉, (2) 𝖔𝖙𝖍𝖊𝖗 𝖒𝖆𝖙𝖙𝖊𝖗𝖘 𝖓𝖔𝖙 𝖈𝖗𝖎𝖒𝖎𝖓𝖆𝖑.[1]

The Rolls of the King's Court or Curia Regis end with the reign of Henry III.,[2] Edward I. having then split up the jurisdiction into three by the constitution of (a) the King's or Queen's Bench, (b) the Common Pleas or Common Bench, and (c) the Exchequer of Pleas.

[1] For criminal matters see next chapter, and for a separate report on the Records of Wales and Chester see 1st Report, pp. 79-122.

[2] Mr. M. M. Bigelow published in 1879, under the title of *Placita Anglo Normannica*, reports of certain law cases, from William I. to Richard I., preserved in historical records, but these are chiefly interesting to the law student, and might have been immensely increased by further research.

Before this subdivision, the Rolls were sometimes styled—

Placita Coram
- Domino Rege in Parliamento suo, *or*
- ,, ,, et consilio suo, *or*
- Concilio Domini Regis, *or*
- Domino Rege et locum suum tenentibus, *or*
- A B justiciar' Angliæ, and so on.

A folio vol. of *selections* from these rolls, from 6 Richard I. to 20 Edward II. was printed in 1811 by the Government, under the title of "Placitorum Abbrevatio."[1] A facsimile of a part is in Nos. xxvi., xxvii., and xxviii. of 1819 Report, and now sells at a low price.

It must be borne in mind that this is a selection, and a small selection indeed only of the vast mass of rolls which are extant; and moreover, that the selection was made by Agarde, who as a rule was on the look out for entries relating to well-known families, and preferred to print them rather than entries which might have a greater historical interest to us now. Very few extracts indeed were given about criminal matters. The whole (?) of the rolls for Richard I. and the 1st year of John were published in 2 vols. 8vo. by Sir F. Palgrave in 1835, under the title of *Rotuli Curiæ Regis*, which sells for £2. 2s. About twenty-five membranes, temp. Richard I., remain in MS., but will be printed by the Pipe Roll Society. When the split was made, the *King's or Queen's Bench* still retained exclusive jurisdiction over criminal proceedings, which were held on what was called its "Crown side," and which are dealt with in my next chapter under Crown Plea Rolls, all other actions being brought on what was called its "Plea side." For an account of the records of the Queen's Bench, see 1800 Report, p. 112; and 1837 Report, p. 131; and as to the more modern records, see 2nd Report, pp. 50-57.

[1] There is a MS. vol. with the same title, which contains extracts temp. Edward I. among *Maynard's MSS.*, Lincoln's Inn, xiii. Three volumes from these and other rolls, were printed in *Prynne's Records* (small folio, 1665-8) which is still a work of great interest and value. Some pedigrees collected from the Coram Rege Rolls by Richard St. George were printed in the *Collectanea Top. et Gen.* i. p. 128, &c., from Rawlinson MSS., Oxford, No. 116.

The Court of *Common Pleas* or Common Bench, whence its rolls were commonly called De Banco Rolls, claimed exclusive jurisdiction over land. These rolls were again eventually subdivided, being separated into the Placita Communia and the Placita Terræ. These De Banco Rolls contain material enough to supply illustrations to every English pedigree, for nearly every family must have had a dispute about land at some time or another, and the pleadings very often give long descents and proofs of the greatest value.

Unluckily, however, the entries on the roll were written down just as they happened to come in, one after the other, without the slightest attempt at arrangement, and the bulk of the rolls is so immense that a lifetime would not suffice to search them. I know it is said that private indexes exist; but they cannot include a hundredth part of the entries, even if so much. For example, there are 151 rolls, containing 102,566 skins, for the single reign of Henry VIII. There are Doggett or Judgment Rolls, which serve as indexes from 1509; and Doggett Books from 29 Charles II. For an account of the details of the modern records of the *Common Pleas*, see 2nd Report, pp. 57-61; General Reports on the records of this court are in the 1800 Report, pp. 120, 127, and 213 b; and 1837 Report, p. 132.

The *Exchequer of Pleas*, the third Court, was supposed to be devoted to litigation arising directly or indirectly from debts due to the Crown, but of late years jurisdiction in general matters was obtained by a fiction by which the plaintiff alleged that *he* was a debtor to the Crown, and that through the defendant not paying him, he in turn was "the less" able to settle with the King, a fiction technically known as "quo minus."

For excellent (but unindexed) Calendars of the Special Commissions of the Exchequer, from Elizabeth, see 38th Report, pp. 1-148; and of the Depositions by Commission, 38th Report, pp. 150-775; 39th Report, pp. 307-531; 40th Report, pp. 1-466; 41st Report, pp. 1-670; and 42nd Report, pp. 1-312, which carries the calendar down to 31 George II. This is probably the most useful calendar yet produced by the Record Office. For details of the modern records of the Exchequer see 2nd Report, pp. 61-69.

An Index to the orders, decrees, and enrolments of this Court, from Edward I. to George III., is said to be in the Inner Temple Library, and no doubt is that which was compiled by Adam Martin, and published by the Inner Temple in 1819, 8vo. pp. 242. It consists of an Index Locorum to entries on the Queen's Remembrancer's side, which struck the compiler, who was an official of the Court, as being interesting,[1] and sells for 6s. or so.

Similarly Additional MSS., British Museum, 4500, 4508, 4531, to 4531, are said to be Exchequer Bills and Answers from Edward I. to 28 Henry VIII., and Additional MS. 9780, is said to be an Index to the Exchequer Decrees, 1-8 James I.

This Court also, in 1 Elizabeth, assumed an equity jurisdiction known as the "Equity side of the Exchequer" or the "*Exchequer Chamber*," a jurisdiction abolished in 5 Victoria.

The indexes to the Equity side at the Record Office are to—

    (*a*) *Depositions*, 1 to 26 Eliz., an excellent modern calendar.

             1 to 11 James I., ditto.

             12 to 22 James I., ditto.

    (*b*) *Decrees*, 1 to 31 Elizabeth. A volume of old calendars, the counties are in the margin, and the arrangement chronological. This is very valuable and useful, and should be indexed.

    (*c*) *Entries and Decrees*, James I. to Charles II. Old indexes, id.

    (*d*) *Decrees*, 10 to 28 Charles I. Old index, counties in the margin.

    (*e*) *Decrees and Orders* to present time. Thirteen small vols. of indexes.

A return as to the records of the Exchequer Chamber is in 1800 Report, p. 215 b.

Very many actions and other matters are enrolled on the

---

[1] It is to the credit of the Court that it also produced so industrious a man as Jones, whose Index to the Originalia and Memoranda Rolls is mentioned hereafter.

rolls of the Exchequer, and to these there are twenty MS. vols. of calendar, with alphabetical indexes in eighteen vols. These are in the Round Room. Unluckily, they are only arranged in the first three letters, and not entirely lexicographically.

I may here mention that from the early part of the reign of Henry VIII. there are what are called Doggett (*q.d.* Docket) Rolls belonging to each of the three Courts,[1] containing short entries of the Pleadings, Judgments, &c., which are infinitely easier to search than the bulky Judgment Rolls themselves. These Doggett Rolls were afterwards turned into books, the Doggett Books of the Exchequer, which are complete from 1 Elizabeth, which is the earliest.

Very few records give more information of a varied character than those of the Court of *Chancery*. An interesting article on them will be found in the *Genealogist*, iv. (n. s.) p. 71, &c., and some specimens of early petitions are printed at length in the prefaces to the Calendars for Elizabeth, published 1827-32, mentioned below.

The series of petitions which commenced proceedings in Chancery commence 17 Richard II., and very many of this reign exist, and though the series is far from perfect until the reign of Henry VI., yet there are many for the intervening period, though none for Henry IV.

That no calendar of these early proceedings has been attempted, and that few facilities are given for their inspection, is a public scandal. The same amount of trouble which has given us such excellent calendars as those of the Exchequer Depositions by Commission, of a much later date, would have rendered these vastly more interesting documents accessible. A bare list of names of plaintiffs and defendants and counties would be a great boon, but it is denied us, and, consequently, this most interesting series of records is practically unused. There

---

[1] Calendars and indexes of the Docket Rolls of the Common Pleas were ordered to be printed (1819 Report, p. 19), but were afterwards postponed, p. 62.

are no less than 39,038 documents prior to Elizabeth, as may be seen by the summary at p. 38 of the 5th Report. They ought at least to be sorted into counties, and rough lists made of names and places.

For the reign of Elizabeth[1] a calendar of proceedings in three vols. folio was published in 1827-32, and is most useful. It is not, however, perfect, for it does not contain all in the MS. calendar, which in its turn is not complete.

For James I. there is a MS. recent *Calendar Locorum*, in three vols. folio, in the Long Room. Topham's index covers the same reign, and is of plaintiffs' names only. Letter A is printed in the *Genealogist*, iv. (n. s.) pp. 79-89.

The MS. "Bill Books" are perfect from the reign of Elizabeth, and give the Christian and surnames of the parties and short account of the Bill, each volume being divided into counties, but unluckily they are divided into six series, owing to the partition of business between the "six clerks."

The indexes to the "Bills and Answers," "Depositions" (from Henry VIII.), and "Decrees," all date from the same period.

"Affidavits," from Charles I. From 1715 to the present time the original entry books of Bills and Answers[2] are kept in the Long Room of the Record Office. There are also six sets of modern calendars of the "Six Clerks," through whom the business was conducted from 1714-1758. Another similar series, lettered "Study matters and Pleadings sorted," from 1758 to 1800, and yet another to the abolition of the Six Clerks in 1842. There is a separate modern calendar in one vol. of "Records found in the Six Clerks' Seats and Studies, when sorted in 1849;" in plain English, of records mislaid through carelessness."[3]

The "Judgment Rolls," containing the decrees from about

---

[1] A similar calendar to the latter part of James I. was compiled by Messrs. Hook and Cossart, *Cooper* i. p. 384. Where is it?

[2] Mr. Phillimore is printing the Office Index to the Chancery Bills and Answers for 1625-1649, in his *Index Library*. See reference in index to this book.

[3] All these last are more easily searched than the original Bill Books.

25 Henry VIII. to 6 George III., exist, but there are very few later entries.—1837 Report, p. 111.

A description of the books of the "Accountant General," from 1726, will be found in the 1837 Report, p. 121 b; and of the "Enrolment Office," 1837 Report, p. 129 b; and of the "Report Office," 1837 Report, p. 124 b; and of the records of the "Six Clerks' Office," 1837 Report, p. 126 a, and 2nd Report, p. 40; 1800 Report, p. 106; and 1837 Report, p. 126 a.

For the Cursitors' Office (the functions of which were referred to the Petty Bag Office in 5 and 6 William IV.) see 2nd Report, p. 43. The functions of the Petty Bag Office were practically to look after the non-litigious business of the Court of Chancery, and are set out at length at p. 68 of Thomas' *Handbook*.

The *Duchy of Lancaster* had an exclusive equity jurisdiction over its tenants, and an excellent calendar to the Pleadings, Surveys, Depositions, &c., from Henry VII. to Mary (two vols. folio) has been published by Government.

It is especially to be noted by young searchers that the Duchy of Lancaster was not confined to Lancashire, *e.g.*, there are very many entries in this work relating to Norfolk.

The Charters of the Duchy of Lancaster formed a separate vol., which was published by Hardy in 1845, and sells for £1.

There is a good calendar of ancient charters belonging to the Duchy in the 31st Report, pp. 1-41; 35th Report, pp. 1-41; and 36th Report, p. 161. Other calendars are (1) of the Chancery Rolls, 32nd Report, pp. 331-365; 33rd Report, pp. 2-42; and 37th Report, pp. 172-179; (2) of the Patent Rolls, 4 Richard II. to 21 Henry VII., 40th Report, p. 521; (3) of the Court Rolls, 43rd Report, pp. 210-362; (4) of the Privy Seals, 43rd Report, p. 363; and (5) of the Ministers' and Receivers' Accounts, 45th Report, p. 1, which is indexed at p. 123.

CHAPTER VI.

Of Criminal Proceedings, &c.[1]

THE Coram Rege Rolls, otherwise known as Crown Plea Rolls, which were King's or Queen's Bench Records of its "Crown side," included the Assize Rolls, Eyre Rolls,[2] Coroners' Rolls, and Gaol Delivery Rolls. Facsimiles of some of these will be found in Nos. xxvi.-xxviii. of the 1819 Report.

As mentioned before, however, the Crown Plea Rolls often contain many entries as to enquiries relating to Crown property, as well as what was afterwards more usually their regular business, viz.. the recording criminal matters. As to this branch see Sir Matthew Hale's *Historia Plac. Corom.* An excellent descriptive analysis of their contents will be found at pp. 138-141 of Bigelow's *History of Procedure,* Macmillan, 8vo. 1880.

The later Crown Plea Rolls, as far as I can make out, present the various crimes, and leave it to the later Gaol Delivery Roll to record the judgment and punishment. Sometimes, however, the Crown Plea and the Gaol Deliveries form different membranes on the same roll.[3]

They are very often in duplicate, and not infrequently in triplicate. Mr. Maitland, in his *Pleas of the Crown for Gloucester* (1884), p. 47, very plausibly suggests that the curious accord in some respects and the strange difference in others of these duplicates, arose from their being taken down in Court by two different scribes.

To this I will add the conjecture that one copy was the record of the Court, as a Court, and the other for the use of the Exchequer, in checking the Sheriff when accounting for fines, &c.

[1] See Stephen's *History of the Criminal Law*, and Mr. Luke Owen Pike's very valuable *History of Crime in England*, 2 vols. 8vo. 1873.

[2] The Justices in Eyre were superseded by the Circuit Judges, and then only became Justices of the Forest.

[3] *E.g.*, C. P., and G. D. Rolls, 5th Edward III., Norfolk.

The Crown Plea Rolls are most interesting as affording evidence of the manners and customs as well as of the crimes of our ancestors. They refer to all sorts of criminal matters, such as trials for murder, theft, and arson; park breaking, horse stealing, sanctuary and its breach, and so on.

It is greatly to be regretted that more of them have not been printed and translated.[1] Speaking under all correction, I think I was the first to print one in English, when in 1869 I published an abstract of the Crown Pleas and Gaol Delivery Roll, held at Norwich and Lynn 5 Edward III. (1332);[2] and when in 1881 I abstracted so much of the Crown Plea Rolls for Norfolk, 34, 41, and 52-3 Henry III., as related to North Erpingham Hundred (*N. A. M.* ii. pp.159-193).

Since then a much more valuable and most excellent edition of the Pleas of the Crown for the County of Gloucester, for 5 Henry III. (1221) was issued (in 1884), by Mr. F. W. Maitland (Macmillan), a work which will long remain a text book for this class of roll.[3]

A few specimens, or rather a short analysis of them, will show what sort of information these rolls contain :—

1. Roger Ayline was found smothered in a marl pit. He was buried without sight of the coroner: the adjoining villages are therefore fined.

2. William, son of Walter de Bromholm, is drowned in a ditch. His mother did not come to the inquest: therefore she is fined, &c.

3. Matilda, daughter of Roger la Weyte, is found drowned in Suthsted, and William Bond falsely presented that she was an Englishwoman : for which he is fined. (It was necessary to prove at each inquest that the deceased person was English, so as to avoid the heavy amercement called Englishry, for killing a Dane or Norman.)

4. Hugh de Hillington appeals Roger fil' Constance, that at, &c., the latter came and struck him on the head with a

---

[1] As remarked before, the extracts from the Crown Plea Rolls, published by the Government, omit the criminal part altogether.

[2] *East Anglian*, iii. pp. 148-153.

[3] The same able editor's edition of Bracton's *Notebook* has just appeared, and is certainly the best work of the kind hitherto published.

hatchet, viz., towards the left side, and made there a wound three thumbs breadth long and two wide. This he offers to prove by his body. Roger denies it, and will defend by his body: therefore let there be a duel between them. Sureties given—a day appointed—Roger comes, but not Hugh. The latter to be taken into custody, and his sureties fined. The jury, however, agree as to Roger's guilt, so he is to be sent to gaol. Afterwards he gets off by a fine of half a mark, secured by another pledge.

For an inventory of all these Crown Plea Rolls see 1837 Report, pp. 22 et seq., and excellent folio descriptive catalogues on the shelves in the Round Room in the Record Office.

I need hardly say there is no calendar or index to the Criminal Rolls (though the Controlment Books and great Doggett Books, 2 B Crown side, serve that purpose to some extent) till we reach the Pye Books, which, for London and Middlesex, begin in 1673, and for other counties in 1660.

For Criminal Papers, Henry VIII. to Charles II., see 1837 Report, p. 79 b; and there are 29 vols. of criminals' petitions from 1687-1780, with a vol. of index, see 1819 Report, p. 363. The reports of the clerks of assize as to their records are in 1800 Report, pp. 211 and 137 a.

In some cities (as in Nottingham[1]) and in most manors minor criminal matters were dealt with by the Court Leet, and are recorded on the Court Rolls.

The Coroners' Rolls (Rotuli Coronatorum) extend from Edward I. to Henry VI.

Nearly every city of note had its own Coroners' Rolls. Those for Norwich begin very early, 48 Henry III., and are the subject of a very interesting paper by the late H. Harrod, in *Norfolk Archæology*, ii. pp. 253-259, which is the model of what such a paper should be.

The Leicester Coroners' Rolls are touched on in a paper by J. Thompson, in the Winchester vol. of the British Archæological Association, pp. 70-81, while H. P. Riley's excellent translations of the London Coroners' Rolls, for 1275-6,

---

[1] *Records of the Borough of Nottingham*, vol. i., Quaritch, 1882.

will be found at pp. 3-17 of his *Memorials of London and London Life*, London, 1868. A report on the Ipswich Coroners' Rolls is in the Historical MSS. Commission 9th Report, p. 226.

Of more important State Trials,[1] details will be found in the *Baga de Secretis*. An inventory and calendar of them will be found in the 3rd Report, p. 213; 4th Report, p. 217; 5th Report, App. II. 2 f; 6th Report, p. 88; and 7th Report, p. 239. Some account of the modern practice of the Court is in the 1800 Report, p. 117 b.

Attainders were entered on the Parliament Rolls, and there was a volume containing a collection of all attainders, restitutions, and redemptions, from 29 Edward III. to the end of Edward IV., among the Tower Records. See Thomas, p. 11.

There is much as to those who rebelled, as to which see temp. Henry VIII., Sims, p. 141, and 1837 Report, pp. 13 b, 14 a. For the Jacobite Trials in 1694 see a volume under that title, published in 1853 by the Chetham Society.

The printed collections of State Trials published should also be consulted, and for late years the *Gentleman's Magazine*, *Annual Register*, and *Newgate Calendar*, often give interesting details, omitted from the more formal reports.

The *Star Chamber* was a court that can hardly be described as either wholly civil or criminal, but it savoured so much of the latter that it may be mentioned here. It was practically re-founded 3 Henry VII., but was found so dangerous a weapon in the King's hands, that it was abolished 16 Charles I. A report on its documents is in the 20th Report, p. 135, and see 24th Report, p. 54,[2] and Burn's *Star Chamber*.

---

[1] An index to the records of State imprisonments in the Tower of London will be found in the 30th Report, pp. 313-53; see also Brayley's *History of the Tower*.

[2] Some other Star Chamber references may be useful. Bill Answers and Depositions, from 3 Henry VII. to 16 Charles I., 1800 Report, p. 39. Orders made in Court of, in temp. Elizabeth, Com. Record MSS. D. d. xi. 81; Records of the Court of Star Chamber, 1603-1617, among MSS. of the Duke of Northumberland, see 3rd Report Historical MSS. Com. p. 154; Notes of Causes in, 17-19 James I., Com. Record MSS. I. i. vi., 51; Treatise on, id., 54, and see L. l. iii., 2 and 3, and iv. 13; Treatise on (in hand of Charles I.) Coxe MSS., xc. (Linc. Inn.) MS. History of, and of its Jurisdiction, 1800 Report, p. 175 b; Modern Practice of, id. p. 117 b.

CHAPTER VII.

State Papers, &c.

Of course it is very hard to define exactly what should come under this head. They are, however, generally considered to be documents relating to public business at home and abroad, not being of that formal class which is to be found on one or other of the regular series of rolls. Of course, this definition will not strictly hold good: since, for example, very many entries on the Pipe, Patent, and Close Rolls are State Papers of the highest class. But let this pass for the present, and let us consider what are the detached or miscellaneous documents which generally go by the above title.

First and foremost come the *Chancery Records in Filaciis*,[1] which contain many royal and other letters, petitions, &c., of the highest interest. Calendars of them will be found in the 4th Rep., p. 140; 5th Rep., p. 61; 6th Rep., p. 88; and 7th Rep., p. 239. There is no printed index, and they unluckily are not in order of date.

There is a Calendar of the Letters from the Ancient Treasury of Receipt (Henry III. to Edward II.) in the 8th Report, App. II., p. 180, and of various letters from the Chapter House (a roll of 33 Edward I.) in the 9th Rep., App. II., p. 246. A Calendar of the more modern Royal Letters (George I., II., and III.) is in 1819 Rep., p. 364.

Again, there are the *Miscellaneous Records of the Chancery*, which contain very many interesting returns, &c.; see 2nd Report. Unluckily, too, there is no index to these.

Rymer's Fœdera is a work of the greatest importance, and was the result of unwearied diligence on the part of its compiler. Roughly speaking it contains nearly every document of real importance relating to the "diplomatic history," as Cooper aptly calls it, of England. For a more lengthy description of the work itself I will refer my reader to Cooper himself (Cooper on *Public*

---

[1] Commonly called the "Chancery Files."

Records, ii. pp. 89-144), though students will do well to get
the excellent Synopsis of it (in English), which Sir T. D. Hardy
published in 1869-73, and which now sells for about 14*s.*

The *Rolls of Parliament* begin from 6 Edward I., and are of
the highest value.[1]

All the "plea" parts of these petitions, &c., from 1278-
1503, were published by Government in 1764, in six vols. folio,
from the beginning to 19 Henry VII., and a very capital
index has since been published; the whole work can be had
for about £3. 15*s.*[2]

Seven vols. of calendar of the *Records of the Privy Council,*
from 1386-1541, were edited by Sir Harris Nicolas in royal
8vo. in 1834-7, and now sell for about £2. 2*s.* The office
books and registers begin 1540, and are kept at the Privy
Council Office, Whitehall, under such restrictions as render it
next to impossible for the student to use them. Why they,
or at least transcripts of the early books, are not sent to the
Record Office no one can say. There are many gaps which
may some day be filled up from old transcripts, *e.g.,* Maynard
MSS. lxxiv. (Linc. Inn Library), covering from 13th August,
1553, to 12th May, 1559, which is, I believe, missing in the
original.[3] The miscellaneous records and papers are by some
supposed to be in unexplored vaults at Whitehall, but I believe
the authorities there deny this.

[1] Copies of some Early Petitions, Edward I., II., and III., are among
Lord Hale's MSS. at Lincoln's Inn Library. See further as to Petitions,
1819 Report, pp. 103 and 563. An Index Nominum is printed in the
34th Report, pp. 2-162, but it is only alphabetical, and not lexicographical;
the more the pity and shame. As to some formerly in the Chapter House,
see 1819 Report, p. 103. There were forty parcels of petitions from Elizabeth
to Charles II. (1819 Report, p. 363), but these, I believe, have been worked
into the Domestic State Papers.

[2] Except two rolls of 12 Edward II., which have since been printed by
Henry Cole in a jumble of records called *Documents relating to English
History.* Some (?) of them were also printed by Ryley in 1661, under the
title of *Placita Parliamentaria.*

[3] The following appeared in *Notes and Queries,* 7th Ser. iv. p. 327, from
the pen of Professor Dasent.

"PRIVY COUNCIL REGISTER.—It may not be generally known to your

The *Journals of Parliament* begin—for the Lords from 1 Henry VIII., and for the Commons from 1 Edward VI., and are printed with elaborate calendars and indexes. Vardon's *General Index*, from 1547-1714 (pub. 1852), will be found useful.

The *Votes* of both houses have been printed from 1681 to 1825.

Many of the *Debates* have been more or less faithfully reproduced by private members who took notes, and by news letters. Good examples of these are the journals of Sir Symond D'Ewes, and various debates published by the Camden Society.

Statutes or Acts of Parliament[1] themselves have been printed again and again. They are Chancery Records, and enrolled on the Statute Rolls, which date from 1278, and on the Parliament Rolls, which date from 1 Richard III. Ruffhead's

readers that the Council Office has preserved, from the year 1540, a record of each meeting of the Privy Council. The first entry in the series sets forth that it shall be the duty of the Clerk of the Council to enter in a book provided for the purpose the business done at each meeting; and from that day until the present date the record has been faithfully kept up. The condition of the volumes and the character of the entries vary, as is to be expected, with each change in the office of the Clerk of the Council. In some cases the rough drafts only remain, whilst in others we find that a more methodical officer has made a fair copy of his daily minutes, apparently year by year. Carelessness and accident, from which no collection of records can be altogether exempt, have, however, caused several gaps in the series, and my object in writing to you is to ask your permission to lay before your readers a list of the *lacunæ*, in the hope that the missing volumes may still be preserved in private libraries, in addition to the few which are known to exist in the Bodleian and British Museum. Many years ago an inquiry of the same nature was the means of bringing to light at least one missing volume, and it is hoped that if any of the others are still in private hands the Council Office may be permitted by their owners to fill the gaps in the series by taking copies of the originals. I may add that this permission has recently been granted by Lord Salisbury, in the case of the fragment of the register of the reign of Queen Mary preserved at Hatfield. The missing volumes of the register are 1543-1545, 1560-1561, 1568-1569, 1583-1585, 1594, 1603-1612, 1646-1648. Nothing is known as to the fate of these volumes beyond a tradition that those from 1603-1612 were destroyed at the fire at Whitehall on Jan. 12, 1618.''

[1] There are good collections of private Acts in the British Museum and the Law Society.

edition is good for early Statutes. The authorized edition is in 11 vols. fo. (1828), and sells for £10. 10s.

The original Acts themselves are preserved in the Parliament Office, which has them from 12 Henry VII. There is a MS. calendar of the Parliament Rolls from Richard III. to 22 Vict., but it is only an inventory of the titles to the various Acts. There are innumerable editions of the Statutes, but that which gives the obsolete and repealed Acts was published in 1828 by the Record Commission (down to Anne) in 11 vols. folio. The new revised edition, under the direction of the Statute Law Committee, is in 15 vols. impl. 8vo. More recently Government has published a Chronological Table and Index to the Statutes, from 20 Henry III., which has run to many editions, and comes down to the present time; but this does not give repealed Acts.

As to an index to the Reports of the Committees of the House, from 1715-1802, see 1819 Report, p. 111. Many of these reports have been printed.

The persecutions of the Roman Catholics in the time of Elizabeth can be traced through the *Recusant Rolls*, 5th Rep., p. 19, and see *Thomas*, p. 143-44, and for later years see pp. 72-3.

Of the period covering the struggle between the King and the Commonwealth there are masses of documents.

One class is the *Parliamentary Surveys*, taken between 1649 and 1653, which relate to sales ordered in 1649 of the honours, manors, and lands belonging to Charles I., his Queen and Prince, and of the fee farm rents formerly payable to the Crown and the Duchies of Lancaster and Cornwall. A calendar of them is in the 7th Report, p. 224, and an inventory, arranged in places, at pp. 59 and 81 of the 8th Report. They are well described in the 1837 Report, pp. 208 b and 392. An index of those at Lambeth will be found at pp. 397 et seq. of the same report. A calendar of such of them as were in the Augmentation Office will be found in the 7th Report, p. 224, and 8th Report, p. 52.

Abstracts of the Accounts of the Committee for Sequestrations for 1644 are in Cam. Univ. MSS. D. d. iii. 67. As to Berks see 1800 Report, p. 196 a; as to the Forfeited Estates Papers, see 5th Report, pp. 97-130.

The *Royalist Composition Papers*, relating to the same period, extend from 1649 to 1660, fill many vols., and contain statements of the estates, ages, and families, &c., of Royalists, the fines paid for their loyalty, and petitions for the release of their property.

There are two series of modern indexes to them, but compiled by someone who obviously could not read the MS. of the period, for they teem with errors.[1] A short calendar of the lords, knights, and gentlemen who compounded for their estates has twice been printed (1655 and 1733).

Another class of Records relating to the same subject and time are the *Plundered Ministers' Accounts*, which are at Lambeth, and extend from 1649 to 1662 (Lambeth Library MS. 1027, and see *Thomas*, p. 145.)

The State Papers in the ordinary acceptance of the word are being calendared in several classes,[2] viz. :—

Domestic.—Home Office, Colonial, Foreign.

Foreign and Domestic.—Treasury, Spanish, Venetian, Scottish, Irish.

For a list of what have been published, and what are in progress, see Appendix V., p. 157.

In the Appendix to the 28th Report, pp. 140-1 is a *most useful* list, arranged both chronologically and alphabetically, of the more important books containing printed State Papers. A calendar of the documents relating to the history of the State Paper Office to 1800 is in the 30th Report, p. 212.

In the Round Room are preserved transcripts from the *Archives of the Simancas*, and the transcripts of *Venetian* and *Vatican Archives*, too, are in the Record Office, and can be obtained if asked for very earnestly. For a list of the former see 45th Report, p. 62, and 46th Report, pp. 337-381.

To bring the lost, stolen, and strayed " State Papers " back to light is the chief object of the Historical MSS. Commission.[3] It

[1] Mr. Phillimore is printing an index to these in his *Index Library;* as to which see the index to this work.

[2] There was an earlier series begun in 1830, of State Papers during the reign of Henry VIII. (5 vols. 8vo., sells for £1. 10s.)

[3] For full lists to date of the collections reported on, see Appendix VI., p. 160.

is too true that former officials, such as Secretaries of State, seem to have considered either that the papers were their private property, or that they would be safer with them than in the office.

Take for example the way in which Cecil seems to have annexed nearly all the papers of value of the Elizabethan Papers, which, under the title of the Cecil Papers, are being calendared and published by the Historical MSS. Commission.

Again, the Carte Papers at Oxford, extending from 1558 downwards, are most valuable. A MS. catalogue exists, and a transcript copy of it is in the Record Office. A rough synopsis of these MSS. is in Cole, vol. ii. pp. 351-8, and some of them (1641-1660) were published in 1739 in two vols., which sell for 5s. A report on them is in the 30th Report, p. 504, and 32nd Report, pp. 1-236. They chiefly relate to Ireland.

The Clarendon State Papers in the Bodleian Library, covering a period from 1523-1627, have been well calendared by the Rev. H. O. Coxe, in 3 vols. 8vo., which sell for about 21s. Another series, bearing the same title, but chiefly relating to the Great Rebellion (from 1621), was published in 3 vols. folio, during the last century, and sell for £2. 2s. or so. For Treaties and Diplomatic Documents see reference under "Treaties" in the Index.

To the vast masses of "lost" papers preserved in the British Museum, the Bodleian, and the Cambridge University Library, in various College Libraries, at Lambeth, and in the numerous private collections, one can only point vaguely, and recommend careful search in the Calendars of the various Institutions, and in the report of the Historical MSS. Commission just mentioned. Then, too, there are the printed collections, calendared in the 28th Report, as just mentioned. Catalogues of nearly all the principal Libraries [1] and Collections are in the British Museum.

Of course, all that relate to wars on land and water also relate to history, and may be considered State Papers, and books like the *Siege of Caerlaverock*, and the Roll of those present at Agincourt (Nicolas) should be consulted for their respective periods.

[1] For a list see Appendix IV., p. 150.

As to the fates of those who, unhappily for themselves, failed to "make history," the reader may consult the *Baga de Secretis*, the *Attainders*, and *State Trials*, detailed in my last chapter.

A List of Estates forfeited temp. Henry VIII., Edward VI., and Elizabeth, is mentioned in 1800 Report, p. 176 b (see also 1819 Report, p. 193); and Special Commissioners as to Forfeited Estates from Charles I., in 1837 Report, p. 117 b.

Similarly, the rebellions of '15 and '45 have a literature of their own. The Records of the Court of Commission on estates forfeited through them consist of over one hundred vols., and a great number of deeds, &c. (Sims, p. 141, and see 1837 Report, p. 706.) For a calendar of names, see Thomas, pp. 382-389; where also will be found a descriptive catalogue of many books relating to these rebellions. In all, it would seem that there are 7471 vols. and bundles ; an Index to the Calendar of which is to be found at pp. 97-130 of the 5th Report.

No one should be without *Historical Notes from* 1509-1714, in 3 vols., by F. S. Thomas. 8vo., 1856, which gives *three* indexes nominum (unluckily not *one*) of the utmost value, to enable the student to fix dates by identifying persons not to be found in ordinary histories, but whose names often occur in undated documents.

For other lists and lives of high dignitaries, see Foss's *Judges*, Dugdale's *Origines Juridicales* (of which the abridged continuation known as *Chronica Juridicialia* will be found handy and useful), Chamberlayne's *Angliæ Notitia*, Beatson's *Political Index*, Hook's *Lives of the Archbishops*, and Campbell's *Lives of the Chancellors*, the last, however, being nearly worthless, owing to its numerous inaccuracies.

Excellent lists and references to Muster Rolls, Army Lists, and other records relating to soldiers of all dates, will be found at pp. 433-439 ; and of sailors, from pp. 439-441 of Sims' *Manual*. For documents bearing on the early English army see Early Army Accounts from Henry III.; 20th Report, p. 112 ; Descriptive Catalogue of Documents, Muster Rolls, &c., 48 Henry III. to 14 Henry VI., 24th Report, p. 32 ; Muster Rolls, 1663-1684, MS. Inventory at the Record Office ; a Catalogue of War Office Papers, 16th Report, p. 23, and a List of Calendars of 24th

War Office Report, pp. 81-84. The names of officers from 1705-1755 are in the *Angliæ Notitia*, and afterwards in the regular Army Lists preserved in the British Museum and elsewhere.

Much as to the *Army* will be found in Thomas' *Handbook*, pp. 10, 35, 161, and 373, and Sims' *Manual*, pp. 81, 436, and 437.[1]

I may mention that the history of many regiments has been separately written. Special enquiry as to these should be made to the editor of any army paper, as the list is too long for insertion here.

As to the *Navy* and matters relating to it, the Admiralty, and so on, I may refer my readers to Thomas' *Handbook*, pp. 8, 58, 231, 337-339, 360-368, 374; and to Sims' *Manual*, pp. 81, 82, 439, 440.[2]

All Summonses to Parliament for the reigns of Edward I. and II. will be found printed *in extenso* in Palgrave's *Parliamentary*

[1] These rough references of my own may as well be given here :—

Placita Exercitus Regis, 24 Edward I. See 1800 Report, p. 38.

Roll of Expenses of Edward I. (30 Edward I.) Id., p. 176 a.

Charges and Expenses of Army, temp. Henry VIII. 1837 Report, p. 13 b.

Military Papers, Henry VIII. to Charles II., 30 folio vols. 1837 Report, p. 79 b.

Collections of Records relating to the Military Defences of England. See Lincoln's Inn MSS.

Musters, temp. Henry VIII. See 1837 Report, p. 14-47 b.

Payments of Ordnance, 5 Henry VIII. Id., p. 13 b.

Accounts of Deliveries of Ordnance and Stores, 1568-82. 1800 Report, p. 172 a.

Accounts of Sir Ralph Sadler. 1800 Report, p. 176 a.

Indentures of War, Edward III. to Henry VII. (5 large folio vols.) 1837 Report, p. 78 b.

[2] The following references from my own note book may also be useful :—

List of Documents as to Navy from Edward III. 20th Report, p. 131.

Report on Admiralty Records. 8th Report, Appendix I. p. 4.

Report on Admiralty Courts Records. 1800 Report, p. 258 b.

Report on Instance Court of, id. p. 304 ; on Prize Court of, id. p. 303.

Treatises on Office of Admiral, vol. xliii. of Lincoln's Inn Library (Hales MSS.)

Many papers as to the Admiralty in library of Magdalen Coll. Cambridge (? through Pepys.) See 1837 Report, p. 338 b.

Writs and Writs of Military Summons, four sumptuous fo. vols. issued by the Government in 1827-34.

The Writs of Summons for Peers are set out in the *Report on the Dignity of a Peer of the Realm*. As to returns of M.P.'s see much in Sims, pp. 154-157, and the Return of 1878 (1213-1702).

Beatson's *Political Index* contains a useful list of all hereditary honours, public offices, and persons in office from the earliest period to 1806, (3 vols. 8vo., sells for about 12s.) Whitely's *Chronological Register* of both houses gives the names of all members from 1708-1807, and sells for about the same price. At p. 74 of the 24th Report is a useful calendar of all Civil Lists from 1698.

Roll as to Protection of Eastern Counties Seaports, 24 Edward I. 2nd Report, p. 57.

Roll entitled "De Superioritate Maris Anglie et Jus Officii Admirallatus in eadem," dated 26 Edward I., found among the Records of the Tower, 1819 Report, p. 188.

Thirty-one volumes of Admiralty matters, by Sir Julius Cæsar, of the Admiralty, are in Lansdowne MSS.

Register of Admiralty of Norfolk in 1536 (Lynn Fisheries, &c.) Linc. Inn MSS., xci. Hall MSS.

Two Rolls as to Navy (5 Edward I. and 10-11 Edward I.) 1837 Report, p. 187 a.

Return of Sheriffs of Norfolk and Suffolk of all Ships in those Counties, 22 Edward I., 1819 Report, p. 188.

Index to all matters relating to the Navy, in 6 vols. of copies of the Parliament Rolls, 5 Edward II. to 19 Henry VII. (? by Pepys) in library of Magdalen Coll. Camb. See 1837 Report, p. 338 b.

Roll of Ships in Norfolk and Suffolk, 14 Edward III. 2nd Report, p. 65 : Lynn, 16 Edward III., id.

Charges of the Navy, temp. Henry VIII. 1807 Report, p. 13 b.

Expenses of building the Grace Dieu. Id.

Account of the King's Ships, temp. Henry VIII. 20th Report, p. 89.

Names of Ships which served in the War against France, 36-38 Henry VIII. Camb. Univ. MSS. D. d. xiii. 25.

Goods of Pirates, temp. Elizabeth. 20th Report, p. 133.

Various Commissions as to Docks, 1714.

Records of the Court of Admiralty—Calendar. See 24th Report, p. 52.

Catalogue of MSS. of the Court of Admiralty. Id., pp. 62-67.

Report on Specimens of early Log Books. 8th Report, pp. 4, 202.

CHAPTER VIII.

Ecclesiastical and Monastic Records.

Of the history of the parish church and its fabric I have treated already in Chapter II. I will now go more fully into the sources of history in other ecclesiastical matters.

Of course Domesday, as usual, is, with the exception of the earlier Chronicles and Anglo-Saxon Charters, the first document of importance to be consulted.

The Clerical Subsidy Rolls are of less interest than the Lay: the detailed lists of them are in the Round Room at the Record Office.

The *Taxatio Ecclesiastica Anglie et Wallie* (or the "Verus Valor"), taken about 1291, when Pope Nicholas IV. granted[1] a tax on all temporal possessions of religious persons to Edward I., in aid of the expenses of a crusade, gives the values of the properties and rents by all ecclesiastical bodies in the various deaneries.

Nothing but amounts and names of places are given, but it is very valuable, as giving a skeleton rent roll of the possessions of priories, &c., at this early period.

It was printed in 1802 by the Government in folio, and sells now for 5s. or 6s.

The *Nonarum Inquisitiones*, or Nonæ Rolls,[2] are just about half a century later (15 Edward III.), and give details of a subsidy of one-ninth granted to the King in aid of his wars. They are specially valuable, as comparing the then value of each

---

[1] A precedent for this grant was that by Innocent IV. to Henry III., of all first-fruits and tenths from 1253, for three years. The taxation under which is known as the Norwich Taxation or Pope Innocent's Valor, or the Vetus Valor. See *Cooper*, i., p. 408. For a facsimile of part of it see 1819 Report, No. xxii.

[2] For an account of them see 1800 Report, p. 151.

benefice with that given in the *Valor Ecclesiasticus*, just described. They also give the value of all properties in the parish belonging to any abbey, priory, or other religious body.

They are also valuable as often giving the names of the jurors, who may be supposed to be the principal men in each parish. For a facsimile of part of them see 1819 Report, No. xxiv.

They were published by the Government in 1807, in folio, and sell for about 9*s.*; but the book is not perfect, only including the counties of Beds, Bucks, Berks, Cambridge, Cornwall, Dorset, Essex, Gloucester, Hereford, Herts, Hunts, Lancaster, Lincoln, Middlesex, Northampton, Stafford, Suffolk, Sussex, Warwick, Wilts, Worcester, Yorks.

The *Valor Ecclesiasticus*, or *Liber Regis*, is a compilation from returns made by commissioners under the Statute 26 Henry VIII., cap. 3, which gave the first-fruits and tenths to the King. For a facsimile of part of it see 1819 Rep., No. xix. It contains surveys of the possessions of nearly every bishopric, abbey, monastery, priory, college, or other ecclesiastical institution throughout England, and was published in six vols. folio by Government.

The best calendar of all those who rose to eminence in the Church or in either university is to be found in Le Neve's *Fasti Ecclesiæ Anglicanæ*, continued by Sir T. D. Hardy down to 1854. The authorities cited by Sims at pp. 418-427 will also be found very valuable.

The most comprehensive account of English *Monastic Establishments*[1] will, of course, be found in Dugdale's, or, as it rightly should be called, Dodsworth's *Monasticon*. The first edition[2] was printed in three vols. folio, 1655-1673, and there have been several others, but the best is the last, by Cayley, Ellis, and Bandinel, nominally in six folio vols., but usually bound in seven. This last edition, though it has a fairly good sort of index rerum et locorum, has practically no index nominum, so that the old edition is still useful. It is very much to be desired that the Index Society should take this matter in hand,

---

[1] Browne Willis' *Mitred Abbeys and Conventual Cathedral Churches* may also be consulted.

[2] This has a very fair index of names, unhappily wanting in later editions.

for it would be of the highest value to genealogists. If not, each county society should have its own entries separately done.

Some counties and dioceses have what may be styled local Monasticons, and among them may be mentioned the *Monasticon Eboracense,* by Burton (York, folio, 1758). Thorpe's *Registrum Roffense* (fo., 1761), Hugo's *Mediæval Nunneries of Somerset and the Diocese of Bath and Wells* (8vo., 1867), and Oliver's *Monasticon Diocesis Exoniensis,* Cornwall and Devon (fo., 1846-1854.)

Taylor's *Index Monasticus* (fo., 1821) is also a very useful book for East Anglian antiquaries, and so is Newcourt's *Repertorium Ecclesiasticæ Londinense* for Londoners; but the latter gives parochial matters only.

All the accounts in the above works may be greatly improved by the very numerous entries relating to these scattered over the Pipe, Patent, and Close Rolls; but of detailed records none are more valuable than the numerous *Chartularies,* or Ledger Books as they are sometimes called. These Chartularies, which contain transcripts of all the charters granting land to the foundation, and often of other documents relating to it, are of the highest interest. Also see pp. 7 and 15.

An attempt at a list of them will be found at pp. 74, &c., of vol. i. of the *Collectanea Topog. et Geneal.;* another by Sir T. Phillipps, which he privately printed in his *Middle Hill Press;* and another at pp. 14-26 of Sims' *Manual,* but this gives only those which are in *public* libraries. A calendar of those at the Record Office will be found in the 8th Report, p. 135; and 20th Report, p. 114. Sometimes in the same volume as the chartulary, but generally separate, is the *Chronicle* of the institution.

Many of these have been printed, for example, the *Chronicon Petroburgense,* by Stapleton (Cam. Soc., 1849), which is specially valuable as containing copies of legal proceedings in which the abbey was involved, and which will serve as precedents for the scholar. The same society has also published *The Chronicle of Jocelin de Brakelond* (Bury St. Edmund's), *The Register of Worcester Priory* (by Hale), and *The Domesday of St. Paul's,* all most ably edited works of the highest interest. An analysis of *The Chartulary of St. Nicholas at Exeter (Coll. Topog. et Geneal.* i.

pp. 60 et seq.). is also worth consulting, as are Hart's *Chartulary of Ramsey Abbey* and Brewer's *Malmesbury Abbey* (Rolls Series.)

These chronicles[1] often give necrologies of the benefactors, and sometimes memoranda as to their burials, their entrance into participation of the benefit of the institution, and so on. In fact, it has been rightly said that they were the precursors of the parish registers. Sometimes they made a separate volume for the *Necrologium*, as may be seen by referring to Dr. Potthast's well known work. *Missals*, noted as belonging to particular abbeys, should be searched for entries in their margins as to persons in the neighbourhood Private missals often contain the dates of the "obits," for whom their owners were bound to pray, *e.g.*, Friar Brackley kept memoranda (circa 1466) of the anniversaries of his employers' relations and ancestors (see *Norf. Ant. Misc.* vol. iii. p. 424.) Of course, the *Mortuary Rolls*, sent round from one monastic establishment to another, must not be forgotten. A good example of one was that of West Dereham, printed by the late J. G. Nichols in the Norwich vol. of the *Transactions of the Archæological Institute*, 1847, p. 99.

As to the fabric, you can consult the *Fabric Rolls* of each cathedral, &c. (see Brown's *Fabric Rolls and Documents of York Minster*, 8vo., York, 1863.)

A good specimen of the Monastic Register is that of Winchcombe, analysed in the *Collectanea Topograph. et Geneal.*, ii. pp. 16-37. Other examples which occur to me are Hutton's *Coucher Book or Chartulary of Whalley Abbey*, 4 vols. (1847-49, Chetham Society), which is very good, and the *Register of St. Osmund* (Rolls Series.)

Of Alien Priories and of their suppression there is a class of records by itself,[2] and so there is of the Knights Templars, the Knights Hospitallers, &c. The possessions of the Knights

[1] Luard published 5 vols. styled *Annales Monastici*, giving the annals of eleven monasteries and priories (which sell for about £1. 5s.)

[2] For list see 20th Report, p. 112, and for extents and bailiffs' account, temp. Edward I. to Richard II., see 1819 Report, pp. 9, 159, 196, and 358. A description of the documents relating to them will also be found in the 1837 Report, p. 159 b.

Templars were seized 1 Edward II., and extents and inquisitions of their manors, churches, and titles exist.[1] See *Cooper,* ii. p. 361, as does an account of their manors given to the Knights of St. John. As to both orders see *Indexes to Bodleian Library.* There are also extents and inquisitions of the property of the Alien Priories taken in the reign of Edward I. and II. and Richard II. (Queen's Remembrancer.) Nearly all these records were transcribed by the Record Commissioners (I believe with the idea of a continuation of the *Monumenta Historica*), and will be found at the British Museum, in three vols. folio.

We now come to a series of records forming a class by themselves, which may well be styled *Dissolution or Suppression*[2] *Records,* as referring solely to the suppression of the monasteries by Henry VIII., and consisting of several classes, such as—

(1) The Acknowledgments of Royal Supremacy taken in 1534. An inventory of these is in the 7th Report, Appendix II., p. 279.

(2) The Deeds of Surrender (Inventory in 8th Report, Appendix II., p. 6.)

(3) The Ministers' Accounts, which contain complete surveys of the possessions of the monasteries, &c., at the Dissolution, made up in annual rolls. The *first* account is the most important one. After 1 Edward VI. they appear on *county* rolls.

(4) The Particulars of Grants, temp. Henry VIII., Edward VI., and Elizabeth. These often give descriptions of the sites of abbeys and monasteries, and always of their properties, value, &c. Facsimile samples will be found in the 9th Report, pp. 148, 149, and there is an Index under the names of purchasers on pp. 148 et seq. of the 9th, and pp. 223 et seq. of the 10th Report.

---

[1] For notes on an inventory of their possessions in 1312, see *Norfolk Archæology.* viii. p. 90; and for a list of their records, 20th Report, p. 98. See also Book of Possessions of the Templars in 1185, 1800 Report, p. 138 a; Book containing account of part of the possessions of the Templars, by Jeffry Fitz Stephen, Master of the Order in 1185 [same MS. ?], 1837 id., p. 157; Description of Documents as to, 1837 id., p. 159 b.; various Rolls as to the possessions of, 1 Edward I., 1837 id., p. 190 b. There were transcripts of the extents of the Templars, &c., made under the old Commission from the Remembrancer of the Exchequer, see 1819 id., p. 37-55.

[2] For a list of documents relating to this class see 1800 Report, pp. 178-180.

MS. indexes locorum exist at the Record Office in four vols. (Division D.) There are also two vols. of MS. calendar of grants, temp. Philip and Mary, and James; and Sir T. Phillipps published an index to the Particulars of Grants, temp. Edward VI. (folio), which sells for 13s.[1]

(5) The Augmentation Office Records contain many documents, such as "Particulars of Leases," "Calendar of Conventual Leases," "Calendar of Fee-farm Rents." For reports on these records see 1800 Report, p. 210 a; 1837 Report, p. 207; and 20th Report of Deputy Keeper, p. 207.

(6) The Chantries. For lists of accounts of their lands, temp. Hen. VIII., see 1800 Report, p. 181 a; and for a valuation, temp. Elizabeth, see 1837 Report, p. 158. Page 207 b of the same report mentions two folio MS. volumes, containing particulars for the sale of their possessions (Henry VIII. to Edward VI.) These are indexed, 24th Report, p. 33; 1819 Report, p. 194. As to the particulars for sale of the chantry lands and rentals of Somerset, Devon, and Cornwall chantries, they are referred to in the 1800 Report, pp. 185 b, 186 a, 189 b. A MS. index of the Chantry Certificates is in the Round Room.

Even an analysis of the "*Bishop's Registries*"[2] would fill a vol., if space could be given. To them were transferred, *en bloc*, as a rule, the most valuable records belonging to the priories, to whose cathedrals the bishops succeeded.

There are also a great number of *Household Rolls and Accounts*, of which a good example is the Roll of Household Expenses of Bishop Swinfeld, 1289-90, by Webb (Cam. Soc., 1853-4), which is very useful for the explanations it gives of the abbreviated accounts for eating and drinking, and other household expenses. The first vol. contains the text; the second, illustrations and notes (the latter being especially valuable and good), and a fairly complete glossary. As to household expenses in the thirteenth and fifteenth centuries, Beriah Botfield's *Manners and*

[1] As to inventories taken in the houses of various friars, in 29 and 30 Henry VIII., see 1837 Report, p. 12.

[2] These must not be confounded with the Bishops' "Registers." One of the latter (Bishop Drokenford's) has just been printed by the Somerset Record Society.

Household Expenses of England (Roxburghe Club, London, 1841) may well be consulted.

As to Bishop's Temporalities, there are three vols. of MS. calendar of particulars about them among the Queen's Remembrancer's papers. A calendar of Bishop's lands, sold between 1647 and 1651, is printed in *Collect. Top. et Geneal.* i. pp. 3 et seq.; and a valuation of the estates of the various bishoprics in 1647 is Rawlinson MS. 240 a. The records of the various cathedrals were reported on in the 1837 Report, pp. 286 et seq., and subsequently in great part by the Historical MSS. Commission. See index, in which references to them will be found.

As to Cathedral Libraries, see Beriah Botfield's *Notes on the Cathedral Libraries of England*, Pickering, 1849.

The history of the *Guilds*[1] has been dealt with very thoroughly by the late Mr. Toulmin Smith in his work,[2] in which he was the first to print any quantity of the most interesting certificates, dated Richard II., as to their foundation. I was, however, the first to bring these certificates prominently into notice, and had previously printed some of them (in conjunction with the late John L'Estrange) in the *Original Papers of the Norfolk and Norwich Archaeological Society* (vii. p. 105), having discovered them in bundles 309 and 310 of the Miscellanea of the Chancery.[3]

There is a good MS. Calendar by Mr. W. D. Selby in the Record Office. The returns of a few counties only have been preserved: Norfolk as usual is very lucky, and so are Lincolnshire and London.

There are in existence many Record or Register Books of important Guilds or Trading Companies, such as St. George's Guild of Norwich (copy *penes me*, and see Hargrave MS. 300), Corpus Christi of Boston (Harleian MS. 4795), Lichfield (Ashmolean MS. 1521). The Book of the Corpus Christi Guild of York has been printed by the Surtees Society. Mr. Oliver

[1] For Letters Patent, incorporating a guild, see Patent Roll 26 Henry VI· part 2, m. 15.

[2] *English Gilds*, Early English Text Society, 1870.

[3] They were stamped for me, as I took them out. I mention this more particularly as the "discovery," *quant. val.*, has been claimed by others. As a matter of fact it was no discovery at all, for the Calendar disclosed their existence, and some are referred to in Palmer's Index, vol. cvi.

has printed the *History of the Holy Trinity Guild of Sleaford*, and the Rev. J. E. Millard has recently printed the book, *Account of the Fraternity of the Holy Ghost, Basingstoke*, 1557-1664. Either will serve as a text book for the enquirer.

Short reference to the goods of the later guilds, at the time of the suppression, will often be found in the Certificates of Church Goods next referred to. Harleian MS. 605 is a *Survey of the Possessions of Gilds and Chantries granted to Edward VI.*

Of great interest to all Churchmen are the numerous *Inventories of Church Goods and Furniture.*

Early inventories, as of the fourteenth century, are scarce. There is a vol. dated 1368 at the Record Office, containing about 800 inventories of the ornaments of all the churches in the Archdeaconry of Norwich. This book was excellently described by the late H. Harrod in *Norfolk Archæology*, v. p. 89. Of the inventories taken in the reign of Edward VI., when the ornaments of the Church were cut down so closely and all superfluities directed to be sold, calendars will be found in the 7th and 9th Reports, pp. 307 and 233 respectively.

Very many have been printed, especially for Norfolk, by the present writer and others, while some counties have had them printed as a whole, *e.g.*, Lincolnshire, by Edward Peacock, F.S.A., *English Church Furniture, consisting of Church Goods destroyed in Lincolnshire Churches, A.D.* 1566, London, 1866, which contains excellent glossaries and indices; and Surrey, by the late Mr. Tyssen-Amherst. See also *ante*, p. 16 and 16 n.

The "Bishop's Registers" give the names of all the clergy presented to all livings, as well as those who presented them. London begins in 1306, and Durham in 1311. Archbishop Peckham's Register has been printed in the Rolls Series, and among other ecclesiastical documents at the Record Office are Indices of Institutions to Benefices extending from 1615 to 1816, which give the names of the patrons and of the clergy presented, with the date of their institution from the King's books, to which there are three vols. of indices. For a catalogue of the Certificate of Institutions at the Record Office see 20th

Report, p. 92 ; and for a list of those enrolled on the Patent
Roll, temp. Charles II., see 46th Report, Appendix I.

The *Liber Decimarum*, which was compiled in 1719, and which
contains a list of benefices, showing the true value of small
livings not exceeding £50 per annum, as they were lately
returned into Her Majesty's Court of Exchequer, in order to
their discharge from payment of first-fruits and tenths.

Books of *Compositions for Tithes*, of which there are very
good modern lists, giving in parallel columns the parishes in
which, and the names of the persons by whom, composition for
tithes were made, and the date.

Though not alphabetical they are easy to search, and extend
from 1536-1659, and are very productive of information, genealogical and otherwise.

For a calendar of the *Tithe Suits enrolled* in the Exchequer
of Pleas, *vide* 2nd Report, p. 250 ; and the excellent calendars
to the Exchequer Commissions and Depositions, in the 38th to
45th Reports, are full of references to tithe suits.

All Gifts for *Charitable and Religious Purposes* since 9th
George II. are enrolled in Chancery, and an excellent index to
them forms a thick vol. by itself in Appendix II., No. 1, to the
32nd Report. A catalogue or index to Commissions and Inquisitions on Lands given in Charity, is to be found in the 1837
Report, p. 120 a, and see id., p. 117 a ; and the Charity
Commissioners' Reports of 1825, &c., give a minute description
of all parish charities.

The literature of the "Recusants," the "Roman Catholics," [1]

[1] The Recusants Roll for Elizabeth is described in the 1837 Report, p. 208
b. Recusants in Sussex in 1650, see 1800 Report, p. 195 a.

For returns as to Papists (Anne and George I.) see 20th Report, p. 47 ;
1800 id., p. 195 a ; and 1837 id., p. 118 b.

Names of the Roman Catholics, Non-jurors of England and Wales, who
refused to take the oaths of King George I., with their titles, places of
abode, parishes, townships, where their lands lay, the names of the tenants,
annual value of them as returned by themselves, collected by Mr. Cosin,
Secretary to the Commissioners of the Forfeited Estates, 8vo., 1745 : a

the "Nonconformists," the "Dissenters;" call them what you will, is large and interesting.

References to very many documents relating to them were collected with great care by Sims, and references to them will be found at pp. 144 et seq. of his *Manual;* and Neal's *History of the Puritans and Protestant Nonconformists* (5 vols. 8vo. 1822). The MSS. in Dr. Williams' Library, mentioned in the 3rd Report (p. 368) of the Historical MSS. Commission, should be consulted.

CHAPTER IX.

Parish Registers, Cemetery Books, General Registry Office, Churchwardens' Books, Inscriptions, &c.

Though the registers of many monastic foundations contain entries as to the families of their founders and others who benefited them, such entries were more usually of deaths[1] than of marriages or births, their primary object being, no doubt, to keep records of the dates on which masses due to the family were to be said or sung. Not infrequently, too, entries were made in the missals of parish churches (*Burn*, p. 11).

No doubt the private chaplains of great families kept other

curious book for the topographer and genealogist, arranged under counties of England and Wales. The reprint of 1862 sells for about 3s. 6d.

Roman Catholics, a List of, in the County of York in 1604; transcribed from the original MS. in the Bodleian Library, and edited, with genealogical notes, by Edward Peacock, F.S.A., 8vo. cloth, 7s. 6d., 1872, printed by Gray, Cathedral Press, Manchester.

Records of the English Province of the Society of Jesus (Burn & Oates), 8 vols. 8vo. These vols. are indispensable to anyone searching for information as to Catholic gentry; see also *The Douay Diary.*

[1] See also the Mortuary Rolls mentioned in the last chapter.

similar necrologies, *e.g.*, that kept by Friar Brackley, in which are recorded the obits of the Pastons and Mawtbys, referred to in my last chapter. Many notes of similar books will be found in *Collect. Top. et Geneal.* i. p. 277. There are, also, several, as mentioned by Burn (p. 10), entered in private books of devotion.[1]

The present system of parochial registers was instituted[2] in consequence of an injunction of Thomas Cromwell in 30 Henry VIII., and ought to be perfect from that date, 1538.[3]

How terribly deficient, however, the series is appears from the returns of the Population Abstract of 1801,[4] which showed that there were, out of something like 11,000 parishes in England, only 812 registers commenced in 1538; while the Abstract of Answers and Returns made in 1834 (sells for about 25s.) shows a still more unsatisfactory state of things, many more having been lost in the interval.

As early as 1597 it had been foreseen that accident or design would often cause the loss of parish registers, and to provide against this an injunction of Elizabeth distinctly provided that the incumbent of each parish should annually send his Bishop a transcript of his year's register. This was improved on by an Act of 1812, which provided that the registrar of the diocese should preserve, arrange, and alphabetically index them in places and surnames. Had these wise rules been obeyed nothing would have been easier than to trace a pedigree. Granted that you knew a man died 1650, aged 42, all you would have had to have done would have been to search the transcript of 1608, or at worst 1607 and 1609, to get the information you wanted. But, probably, no injunction was

[1] The best known works on the subject are: Bigland's *Observations on Parish Registers*, 1766; *Origines Genealogicæ*, by Stacey Grimaldi; Burn's *History of Parish Registers*; *Parish Registers*, by R. E. Chester-Waters; *The Preservation of Parish Registers*, by T. P. Taswell-Langmead, with preface by W. C. Borlase, M.P.

[2] This is in England only.

[3] *Burn* (p. 12) notes eight registers which have entries earlier than 1538, viz., from 1528 downwards; but these seem clearly unauthorised entries by too eager clergymen anxious to record the births, &c., of patrons and their families.

[4] Published 1802; sells for about 21s.

more completely set on one side and broken. Early transcripts are simply conspicuous by their absence, and those of the eighteenth century are most imperfect, and in nearly every diocese are left in the utmost neglect and confusion.[1]

What arguments there *may* be for the parochial clergy retaining their old Parish Registers (often to be left loose in a wooden church chest or on the shelves of a parsonage library, to be sold or removed by the executors of the outgoing clergyman with his books), there can literally be none for the retention by the bishop's officials of these transcripts, the collection of which they have throughout neglected so scandalously, while they have taken no care whatever of those that have been sent them by conscientious clergymen. If these were sent to the Public Record Office they could be sorted and flatted with little trouble, and be made immediately accessible to many to whom the original registers are sealed books. There would be no loss of revenue worth speaking of to the authorities, for I was told not long ago that about £1 a year was the average of fees received in one Bishop's Registry for looking at them; 3s. 4d. a year being the ordinary search fee.

As to the original old registers (say before 1754) themselves, no sane man who is accustomed to searching them can doubt that the proper place for them would be either the Public Record Office in London, or some Diocesan Fire-proof Registry, as suggested by Mr. J. S. Burn.[2] The question of fees might be reserved, and, if thought necessary, the clergy of each parish could be credited with the amount collected by the head office. The fact of these being at one office and accessible without journeys or inconvenience, would treble at least the fees now received.

---

[1] Mr. Blaydes is issuing a collection of several thousand entries from the Bishop's transcripts of Bedfordshire, 1603-1700; the subscription being £1. 1s.

[2] The Archdeacons and Rural Deans of Lincoln sent out circulars to the clergy of the diocese, with the view of opposing Mr. Borlase's bill (originally introduced 1882), and issued a series of questions as to the state of the registers: with what results I do not know.

Before quitting the subject of copies I must draw attention to the excellent work done by the Harleian Society and others in printing transcripts of registers.[1] The Harleian Society[2] has already printed many, and especially of Westminster Abbey and Canterbury Cathedral, and of many London parishes, and has the following in hand :—

St. George, Hanover Square (marriages)	Durham Cathedral
	Bath Abbey Church
Christ Church, Newgate Street	St. Sepulchre's, Lambeth.

A very complete list of all parish registers, which have been printed in whole or in part, is given by Dr. Marshall on pp. 194-201 of the *Genealogist*, second series, vol. ii. To these may be added W. J. C. Moen's Register of the Dutch Church, Austin Friars, 1884; and his Registers of the Dutch Church at Norwich, 1887.

The Rev. A. W. C. Hallen has printed St. Botolph, Bishopgate, London (to 1628), and is going on; also St. Mary Woolnoth and St. Mary Woolchurch, Strand, London (1538-1760.)

Mr. F. A. Crisp has recently privately printed,[3] besides those mentioned in Dr. Marshall's list, *ut supra*,—

Ongar	Essex.	Culpho	Suffolk.
Stifford	,,	Ellough	,,
Staines	Middlesex.	Frostenden	,,
Brundish	Suffolk.	Kelsale	,,
Carlton	,,	Tannington	,,
Chillesford	,,		

The Rev. A. G. Legge is printing the register of
North Elmham . . . . Norfolk.

Sir T. Phillipps :—
Durnford . . . . . . Wilts
Somerset House Chapel . Middlesex.

Mr. J. A. Squires is printing
Wandsworth . . . . . Surrey.

---

[1] I shall be most thankful for descriptions of all printed registers, so that I may make the list perfect in a future edition, if this book ever runs to one.

[2] Application for membership to Mitchell & Hughes, 140, Wardour Street, W.

[3] For detailed list and prices of them apply to Mr. Crisp, Grove Park, Denmark Hill, S.E.

Many parish registers have been transcribed, and especially must the Rev. F. Procter of Witton be honoured, for he has done the following :—

Antingham . .	Norfolk.	Ridlington . .	Norfolk.
Bacton . .	,,	Sloley	,,
Catfield . . .	,,	Somerton, East & West	,,
Crostwight . .	,,	Sutton . . .	,,
Dilham . . .	,,	Thorpe Market .	,,
East Ruston . .	,,	Trunch . . .	,,
Happisburgh .	,,	Walcot . . .	,,
Honing . . .	,,	Winterton . .	,,
Horsey . . .	,,	Witton by Walsham	,,
Knapton . . .	,,	Walsham, North	,,
Lessingham (with		Walsham, South	,,
Hempstead and Eccles)	,,	Westwick . .	,,
Mundesley . .	,,	Worstead . .	,,
Palling(with Waxham)	,,		

Mr. T. J. Cooper has copied and given to the Rev. F. Procter,
        Paston . . . . . Norfolk.

The Rev. H. T. Griffiths has copied
        Felmingham . . . Norfolk.
        Smallburgh   ,,

Mr. W. Rye has copied, and is printing
        Bircham Newton . . Norfolk.

Mr. R. G. Rice has copied
        Mitcham . . . . . Surrey.

Mr. A. R. Bax has copied
        Horley . . . . . Surrey.
        Pleystone   ,,
        Friends' Register at Capel   ,,

The late Archdeacon Thorp had transcripts made of
        Alston (2) . . . . Cumberland.

Mr. Lucas of Leeds transcribed
        Leeds . . . . . . Yorks.

As, however, the parish registers are still left with the parsons, I will give a few directions as to searching them.

The fees payable by searchers are 1s. for the first, and 6d. for every subsequent year searched.

This of course includes a search of all the entries of baptisms, marriages, and deaths in any one year; but there are some clergymen who, too eager to increase their emoluments, try to charge 6d. each for the baptisms, the marriages, and the deaths, and I was once asked to pay £12, for an antiquarian search extending over one hundred and sixty years, on this ground, but successfully resisted the claim.

To the honour of clergymen in general, however, I must say, such instances are extremely rare, and in the great majority of cases, if the searcher do not unduly take up the time of the custodian, he is seldom charged anything, if his search partakes of an antiquarian or general character. Within the last one hundred years, however, it is customary for the full fees to be charged, and this, I think, is but fair. For long searches a nominal sum of 10s. 6d. or £1. 1s. is sometimes charged, and being much less than the legal fee, should always be cheerfully paid.

For every official extract made and certified by the clergyman a further sum of 2s. 7d. should be paid, which includes the 1d. inland revenue stamp. To render this certificate more easily provable, and especially in Chancery, the person attesting the certificate should subscribe himself "A B, Rector, or Vicar, or Curate of the above parish of X." Certificates signed A B, Rector, &c., only, do not prove themselves as it is technically called, for A B may be the rector of some other parish, and not the proper custodian of the register. A certificate attested by the parish clerk or a churchwarden (I have seen such) is valueless.[1]

In writing to a clergyman at a distance for a copy of an entry—the exact date of which you know—you should forward 3s. 8d., for he is entitled to 1s. for the search fee, as well as the 2s. 7d. certificate fee, the odd 1d. being for return postage. Unless you are sure of his surname, you had better address him "The Officiating Minister, X," for if you direct your letter to an incumbent who has left the neighbourhood, your

---

[1] For directions as to framing and proving a "Chancery" pedigree, see Mr. Burney's admirable chapter in his edition of *Daniell's Chancery Forms*, which is a model of what such a work should be.

letter and stamps may be lost travelling after him, and in any case will only give him and yourself useless trouble.

Should you want a general search made for any of your name, it will be as well to offer a certain fee, and not to leave matters uncertain.

Some clergymen, if good natured, and with a small parish, will often look through their registers, on the understanding that you will take and pay for official copies of all entries they find relating to the name you wish searched for.

If the clergyman is a member of the same archæological society as yourself, you should mention the fact in your letter, in which case, if the search is not purely a business one, he will of course not make any charge.

You should remember that if the person to be searched for gained a scholarship at a great public school, or has been at Oxford or Cambridge, a certificate of his baptism will often be found among the college records. Similarly, if there was a policy on his life, a certificate is often lodged, to prove his age, at the office; but whether kept beyond a certain time I do not know. Still, I once found this useful in a pedigree case.

The non-parochial registers, i.e., those kept by the various Dissenters, including Quakers, are now preserved in the General Register Office, Somerset House,[1] under two commissions in 1836 and 1857.

They are not of course indexed; but catalogues of them were published in 1841, 1858, and 1859. They include French, Walloon, German, Dutch, and Swiss church registers. Antiquaries and others are apt to overlook the importance of these, forgetting the period to which they sometimes go back.

Among other private registers was one of great importance, known as the register, formerly kept at Dr. Williams' Library, Red Cross Street, but now at the General Register Office, Somerset House. A catalogue of the library was printed in 8vo. in 1841, which contains entries of all denominations of Dissenters from all parts of England, beginning 1715.

The baptismal entries give the names of the *mother's* father and

---

[1] As to these see Appendix.

mother, and are very voluminous and valuable, being well indexed.

Entries of the marriages of Roman Catholics and Dissenters are often to be found in the parish registers, for until 1806 no marriage was legal, whatever the denominations of the persons married, unless it took place in the parish church—a rule removed by the Marriage Act, 6 and 7 William IV., cap. 85.

Should the parish registers and dissenting registers not contain the entries wanted, the searcher should remember that there were such unauthorized marriage registers as those of Gretna Green and the Fleet.[1]

In 1836 the General Register Office was instituted for England, and from the 1st July, 1837, all births,[2] marriages, and deaths are recorded in quarterly volumes, with perfect lexicographical indexes, which are kept at Somerset House, and which can be consulted on payment of 1s. This allows the searcher to look at three entries, but not to take notes. A "general" search costs £1. 1s., but the authorities do not seem very clear what a "general search" means, and I apprehend that it would not authorize the taking of notes of all entries relating to any one surname which occurs in the indexes. For a list of the various non-parochial and other registers see Appendix II., p. 146.

Of late years the quarterly indexes have been printed, which vastly facilitates searches. Why duplicate copies of these printed indexes are not placed in the Record Office, the British Museum, the Law Institution, and most other large libraries, it is hard to say. They would greatly ease the work of the office, and allow much to be done by correspondence, thus saving searchers who want official copies of certificates from having to wait while they are being copied.

[1] As to these consult Burn's "Fleet Registers," 8vo., 1833. For some entries from the register of Duke Street Chapel, Westminster (from 1709), the Rolls Chapel, St. John's Chapel, Bedford Row, and Wheeler Chapel. see *Coll. Top. et Gen.* iii. p. 381; and Gray's Inn and Knightsbridge Chapels, *Id.*, vol. iv. pp. 157 and 162.

[2] It is to the credit of the Heralds' College that in 1747 they tried to establish a "general register" of births (Burn, p. 77), but it soon died out for want of publicity.

The Scotch Registry does not begin till 1854,[1] and the Irish till 1864. Indeed, Mr. Chester Waters points out (p. 9) that "until 1st January, 1864, the births and deaths of the entire population of Ireland, and the marriages of the Catholic majority, were suffered to remain wholly unregistered." Of course he means by Government, for the parochial clergy, Catholic and Protestant alike, did as a rule keep registers, but not well. The earlier registers of the late Established Church of Ireland are in the P. R. O. Dublin. The *lacunæ*, owing to wanton destruction by neglect, fire, &c., are however very great.

The pig-headed obstinacy of the authorities of the Bank of England will not[2] be satisfied with death certificates, and still renders it necessary, in many cases, to supply burial certificates, *especially when one has to "make a man dead" in respect of Government Stock.* It seems beyond the mental capacity of those who manage the transfer office here to understand that if a man's death is well and truly proved it must be a matter of indifference where and when he was buried, or indeed whether he was buried at all. I very much doubt the bank's right[3] to insist on a burial certificate if sufficient proof is tendered them of the death, and the rule often causes great hardship, for a family may often know when and where a trustee died, but after the interval of years be quite unable to find out where he was buried.

This, too, is not all, for the person making the declaration identifying the certificate often had[2] to swear that he personally examined the certificate with the entry in the burial register. Not long ago I, unluckily, had to make one, and to qualify myself for the declaration wasted about six hours looking up the registers of cemeteries at Lee.

[1] The Scotch Parish Registers begin 1550, but were very badly kept, and many are missing.

[2] *Since the above was in type these bad old rules have both been abolished.*

[3] The bank's "right" to make a man coming with a bank note and asking for gold put his name and address on its back, was insisted on for years, till, as the story goes, the cashier happened to tell a solicitor, even more brusquely and rudely than usual, to endorse the note. That individual walked out without a word, and at once successfully sued the bank on its dishonored promissory note. I hope this tale is true, if not it is *ben trovato* anyway.

G

At Appendix III., p. 148, I have given a list of the chief Metropolitan Cemeteries with their addresses.

What the parish and other registers often fail to show may sometimes be gathered from the Monumental Inscriptions;[1] and for the higher grades of society, in earlier times, from the "Funeral Certificates," preserved at the Heralds' College, of which samples will be found in the *Lancashire Funeral Certificates* by King and Canon Raines, Chetham Society, 1869.

CHAPTER X.

Fiscal Records, the Subsidy Rolls, &c.

PROBABLY no class of records is more valuable, both to the genealogist and to the topographer, than the Subsidy Rolls;[2] which are records of the Court of Exchequer.

To the former, they not only often prove of great use, as containing the only record of the existence of persons not of knightly or gentle rank;[3] but help, by giving the locality where they lived, to give some clue as to what part of the country they sprang from.

To the topographer they are of the greatest interest, as shewing the relative importance of different places at different times, and often affording an insight as to what were the trades and occupations of its inhabitants.

Again, to the student of surnames, the immense number of lists of names so conveniently preserved in columns for search and noting are most valuable.

Of the clerical subsidies, which are in another series, and

---

[1] Printed inscriptions there are in legions, as mentioned at p. 7. A collected index to them is much wanted.

[2] Sims, p. 45, falls into a strange error in saying they contain the supplies to the King from his tenants *in capite*.

[3] For them, see the Knights' fees, &c., p. 31.

which are calendared separately, I have spoken elsewhere, and will now say a few words as to the "Lay Subsidies."

These have been very carefully calendared, and the bound "descriptive" slips forming the calendar are divided into counties, and are to be found in the Round Room at the Record Office.[1]

These calendars also give the Hundreds to which each document refers.

Calendars of the Rolls are to be found in the 2nd, 3rd, 4th, and 5th Reports, and a detailed list of the Exchequer Series in the 20th Report, p. 138.

Some rolls begin as early as the reign of Henry III., but most counties have none before about the middle of the reign of Edward I.

There is generally a fine roll about 1 Edward III.; that for Norfolk, though imperfect still, contains seventy-two long skins written on both sides in double columns, and giving the names of certainly 37,000 persons, with the sums at which they were rated, arranged under the villages and towns where they resided.

The return for the Poll Tax of 2 Richard II. (1379) is also usually perfect. That for the West Riding of York was published in 1882 by the Yorkshire Archaeological and Topographical Association (Bradbury and Co.), and cannot contain fewer than 25,000 names. Unluckily, the Association did not think it worth while to print an index, so its value is minimized, at all events for genealogical collectors.

Very few Subsidy Rolls have been printed, and the only complete series for any Hundred is, I think, that printed by me in my *Rough Notes for a History of North Erpingham,*" part ii. which has all the Norfolk Rolls from 1 Edward III. to 24 Charles II. Sir Thos. Phillipps lithographed the Subsidy Roll for Wilts for 7 Edward II., which now sells for 14s.

---

[1] Very many of these Rolls only contain the names of the collectors and particulars of the amounts received, and it is very aggravating to have these commonly valueless documents brought down after a long wait. There is nothing to shew this in the calendar, but a shrewd guess can often be given by noticing the number of membranes given in the calendar. If there are only one or two, or if the roll is headed "Particule Compoti," the document should not, as a rule, be written for, as no names will be given.

The Hearth Tax Rolls are also very full as a rule, and begin 14 Charles II., but were soon abolished—1 William and Mary.

To those who are interested in tracing the component parts of the present English race, there are many "Alien Subsidies" and other documents in this series which will interest them, *e.g.*, "$\frac{140}{18}$ Roll of Inquisitions, showing the names of foreigners, &c., in the hundred of Earsham, chargeable to the Alien Subsidy."

The records of the Treasury side of the Exchequer are most interesting. A rough report on them will be found in the 7th Report, Appendix II., and a statement of receipts and issues from 1625-1699 in the 7th Report, Appendix II; but the student should consult Thomas' *Ancient Exchequer of England*, 8vo. 1848, which sells for about 6s. The documents relating to the office of the Queen's Remembrancer of the Exchequer are a class by themselves, and are described in 1837 Report, pp. 143 and 194. The miscellaneous records of the Queen's Remembrancer are calendared in the 40th Report, pp. 467-479.

The *Custom Rolls* form another series, and are of the highest value to all who are interested in the growth of English trade. The text book (and an excellent one too) on this subject is Mr. Hubert Hall's *History of the Custom Revenue in England*, 2 vols. 8vo. 1885. For a list of rolls relating to the collection of customs at the various ports, from Edward I., see 20th Report, pp. 115 and 116, and 1837 Report, pp. 185 b-189.

Among other fiscal documents are the *Pell Records*, which comprise Liberate Rolls, Issue Rolls, Entry Books, and recording payments made out of the King's Revenue, and are very interesting. Three volumes of extracts from them, viz., (i) extracts from the Liberate and Issue Rolls, Henry III. to Henry VI.; (ii) The Issue Roll for 44 Edward III., 1370; (iii) Extracts from the Entry Books of James I.; were published by the late Frederick Devon in 1835-7, the first being issued under the title of *Devon's Issues of the Exchequer*, and would sell for about £1. 1s.

Specimens of a short tabular calendar of the Liberate Rolls will be found in the 1837 Report, p. 74.

For reservation of income, &c., to the Crown, see the Originalia Rolls mentioned in Chapter XIII., p. 96.

CHAPTER XI.

The Descent of Land. Inquisitions Post Mortem. Proofs of Age. Wills and Administrations.

THERE can be no doubt that, besides the various presentments of fines due to the King on heirs succeeding to their estates and not being knighted, which we find on the Crown Plea Rolls, there were from early times separate inquisitions taken as soon as any one of importance died, to ascertain of what he died possessed, and who was his heir. Probably the findings on these inquisitions were communicated to the authorities who attended the Court, and made up the roll, and were the origin of the entries in question.

Whether these inquisitions were taken on each death, or periodically, we do not now know.

Traces of inquisitions of the latter class will be found in the roll printed from Harleian MS. 624, which was printed by Stacey Grimaldi in 1830, under the title of *Rotuli de Dominabus et Pueris et Puellis*. The date of this roll is 1185. It is said by D'Ewes to be collected from the Pipe Roll of 31 Henry II. (see 1837 Report, p. 181), and was long missing, but has now been found among the Miscellanea of the Exchequer.

It gives returns as to the heirs and their ages, and the value of their lands, for twelve counties, viz., Beds, Bucks, Cambridge, Essex, Herts, Hunts, Lincoln, Middlesex, Norfolk, Northampton, Rutland, Suffolk; but of course only refers to the marriageable ladies who were then in the King's gift and the boys and girls in ward.

Of the *Inquisitions Post Mortem* [1] there are two series, [2] which were certainly taken separately for each individual, one of the

---

[1] For a form see Appendix I., p. 139. There are also many entries relating to heirships on the Fine Rolls of the Chancery, q. v.

[2] Besides that of the Court of Wards and Liveries mentioned hereafter.

Chancery, and one of the Exchequer; and one series often serves to fill up *lacunæ* or to explain illegible documents in the other. The Chancery series begins nominally 2 Henry III. (1217), but really eighteen years later, and extends to 20 Charles I. (1644).

The Exchequer series begins Edward I. A calendar of them from Henry VII. to Elizabeth will be found in the 10th Report, pp. 2-222, to which there is a MS. index in the Record Office.

It has been said that transcripts were made, under the old commission, of all inquisitions from Edward III. to Elizabeth (see 1819 Report, p. 39), which would be very useful to those who cannot read the old hand; but I believe the statement was only one of those pleasing fictions which abound in the old reports.

When the tenant *in capite* died, the King at once sent down a writ to the Escheator who was appointed for each county, and was so called from having to look after the escheats or fines due to the King. This writ, which was called a *diem clausit extremum*,[1] directed the Escheator to call a jury, who were to enquire of what land the dead man died seized, and by what rent he held it, and who and how old was his heir.

The writ and the return were sent back together into Chancery, where they were filed and are now consultable.

A Calendar to this (the Chancery) series exists from the beginning in 1217 to 2 Richard III. (1484) with an appendix of some inquisitions as late as James I.[2] Unluckily, this calendar is extremely scanty and incorrect,[3] and does not give the heirs, but has long indexes of names and places.

In 1865 another series of Calendars was commenced, two royal 8vo. vols. being published for the Government by Mr. W. H. Roberts, under the title of *Calendarium Genealogicum* (sells for £1.)

---

[1] From its form, which ran as shown in Appendix I., p. 139.

[2] The Appendix contains references to about 3,000 Inquisitions long thought lost, but found just in time to be included out of their order! See *Cooper* i., p. 333 n.

[3] A copy corrected and annotated in MS. will be found in the Round Search Room at the Record Office, where there are also 9 vols. of lists of the Chancery Inquisitions, Henry VII. to Charles I.

It covered the reigns of Henry III. and Edward I. only, and is conceived on opposite lines to the old calendar, for it gives detailed particulars as to the finding of the heir, but omits all reference (except to the county) of the property. This is a great pity, for though a searcher can get the substance of any inquisitions by combining the entries in the two calendars, it gives him needless trouble.

The publication of Roberts' Calendar was stopped, but a few years of Edward II. were printed in the first Appendix of the 32nd Report, and Mr. Vincent [1] is slowly going on with it in the columns of the *Genealogist*.[2]

From 32 Henry VIII. transcripts of the Chancery Inquisitions were sent into the Court of Wards and Liveries,[3] which was instituted in that year (1540); but this Court was practically extinct during the Commonwealth, and was abolished by the statute 12 Charles II., cap. 24 (1660).

The best description of the Records of this Court will be found

[1] In 1877 Mr. Vincent put forward a proposal to print an Index to the Inquisitions during the Tudor period, being a Handbook to the Official Calendars used in the Public Record Office, with the names arranged alphabetically, and grouped under the several counties, followed by a general index; but, unluckily, what would have been a most valuable work was not proceeded with for want of subscribers.

[2] For Lancashire and Norfolk Mr. Selby has published Calendars of the later Chancery and Exchequer and the Court of Wards Inquisitions in his *Lancashire and Cheshire Records*, and his *Norfolk Records*.

[3] Duplicates of these, being the Feodaries Certificates of Inquisitions, were among the Rolls Chapel Records, and are now known as Enrolments of Escheators' Accounts. They were made up for each county every two or three years, and give very nearly the same information as the Inquisitions. A Calendar of these Escheators' Accounts, from Henry VII. to James I., is printed in the second Appendix to the 10th Report, pp. 2-223, and there is a MS. Index in the Long Room at the Record Office. There is an excellent paper on Escheators' Accounts by the Rev. J. Hunter in the 1st Report, pp. 139-142. For Inventories of the Records of the Court of Wards and Liveries, see 4th Report, pp. 81-98, and of the deeds, &c., belonging to the various wards, and left in the Office, see 6th Report, pp. 1-87. A Calendar of the MSS. of the Court is in the 24th Report, p. 54. As to proceedings in the Court of Wards and Liveries, 1540-1660, see 1800 Report, p. 39 b.

on pp. 176-177, of Mr. Walford D. Selby's *Lancashire and Cheshire Records*.

A rough Calendar of the Inquisitions of the Court will be found in vol. civ. of Palmer's Indexes (Record Office).

Thomas Cole collected from the Records of the Court of Wards a great deal of very valuable matter, now forming Harleian MSS. 410, 411, 756, 757, 758, 759, 760, and beginning 32 Henry VIII. An Index to this collection was printed by Sir Thomas Phillipps, and is of very great use.

Probably no class of records has had more extracts taken from, or abstracts made of, them than these Inquisitions Post Mortem. Nearly 4 pp. of Sims' *Manual* (125-8) are taken up with a detailed catalogue of some of them, and the searcher interested in any one county should, by all means, refer first to this, and see whether any one has already been collecting what he wants.

Of published County Calendars there are not many. Phillipps' Index to Cole's collection is mentioned above.[1]

Calendars of the later Inquisitions for Lancashire and Cheshire and for Norfolk, have been printed by Mr. Selby; and the Yorkshire Association have printed one from James I. to Charles I.

Besides the regular series of Inquisitions Post Mortem, it must not be forgotten that there are very many relating to the tenants of the Duchy of Lancaster, whose officials exercised a separate jurisdiction over them. There are three folio volumes of printed Calendars of (*i.a.*) these Inquisitions, and if the searcher is interested in counties in which the Duchy had possessions, he must by no means omit to search them.

When by the Inquisitions Post Mortem it was found that the heir was the King's " ward," his rents were received by the King during his minority or sold by him to some purchaser.

Naturally, therefore, the grip of the King and his receivers on his income was not relaxed until it was proved up to the hilt that

[1] Extracts from some of these for Middlesex are printed in the *Top. et Gen.*, i. pp. 330 and 520; and for Somerset and Dorset in *id.* ii. p. 48; while for a Calendar of Lancashire Inquisitions, Richard II. to Elizabeth, see 39th Report, p. 533.

such minority was really over. This necessitated a second inquisition taken on behalf of the minor, on a writ technically called the "proof of age," on which a jury enquired into, and found the age of the heir.

Apart from their genealogical value, they are highly interesting from the curious events related by the different witnesses as reasons why they remember the birth of the child as to whose age they are deposing.

The late W. D. Cooper printed in vol. xii. of the *Sussex Archæological Collections* an excellent paper dealing with the proofs of age of Sussex families, from the Introduction to which I extract the following descriptive notes:—

"Occasionally the witnesses were present at the baptism, and held a lighted wax candle or torch at the font ; sometimes they saw the child taken to the church, or were informed by the chaplain or servants that the ceremony had been performed. Some witnesses recollect the day, because births and deaths happened in their own families, or, what seemed to fix most distinctly the memory, the witness's own marriage, exemplifying our proverbial East Sussex saying, "he has got a wife, and he knows it ;" one remembers it from the unwilling marriage of his own mother, with whom her future husband went to law, "and it was adjudged by law that he should have her to wife against her will, and he married her that year ;" others remember it from descents, purchases, or grants of land or woods; some because of high winds and the unroofing of houses; or the placing of a new paling round the neighbour's close; or the erection of new houses and stables ; others because the godfathers and godmothers put up their horses, or eat and drank at their houses; one man fetches the midwife ; another brings the godfathers and godmothers; and a third learns the fact from them : institutions to livings, admittances to religious houses as monks, and apprenticeships, all help the recollection.

"Misfortunes (the middle plague amongst them) enter largely into the refreshers of the memory. At Hurstmonceux a rather lawless state of things existed; for whilst on the baptism of one Fiennes, the child's father, the lord of the place, went to the witness's house, and wished to beat him ; the same lord distrained for four chickens in arrear from another tenant, and when, twenty-seven years afterwards, the grandchild was baptized, sixty large eels were stolen from the father's nets. One man gave himself the gratification of walking to Bosham, to look at an acquaintance who had hanged himself; another broke his arm by a fall from a load of hay; one more broke his leg against a ladder, "which he shall never forget;" and, worst of all, poor John Stryveling lost his man-servant on the christening-day, taken by the French enemies at Chidham, and carried to Harfleur."

Unluckily, these proofs of age are not numerous. A list of them is printed in the 3rd Report of the Deputy Keeper, p. 202, and 4th Report, p. 131. The new references to these documents are given in the interleaved copy of the folio Calendar of Chancery Inquisitions Post Mortem, kept in the Round Search Room.

These proofs of age were entered in the Queen's Bench Rolls (*Thomas*, p. 112), and it would be as well if a calendar of these enrolments could be made, but I fear the labour would be too great.

Until 32 Henry VIII., a man, however rich in lands he might have been, had no power to make his "will" and leave away a single acre from his heir.

His *Testament*, dealing solely with his money and other personal property, he could have made, and generally did so in some permutation or combination of the forms in Appendix I., p. 139.

When he died the document was proved, and a certificate or "probate act" (see Appendix I., p. 141) was annexed to the parchment transcript of the will.[1]

Sometimes, however, though he never signed a will (signature by the way, was unnecessary as far as money or other personal property was concerned, till 1838), or even had it reduced to writing, he made what is called a *nuncupative will* or testament on his death-bed.[2]

---

[1] It is most amusing how nearly every novelist and playwright to this day, when, in due course of events, the original lost will which is to restore the wronged persons to affluence is discovered, describes it as "parchment." I never saw an original will on parchment, but the transcript or probate is always copied on it.

[2] For specimen see Appendix I., p. 142, and 2nd Report, Historical MSS. Commission, p. 77.

If he died without written or spoken will, then *administration* to his estate and effects was granted in early days to the Ordinary, but later on to his next of kin.

One need not dilate on the extreme value of all the foregoing documents. No other class of records gives so much information as to the relations, the position, and the class of life of the person who made it. No other series is more interesting, and none, I regret to say, more inaccessible than they.

Certainly, the early wills, formerly at Doctors' Commons, which begin about 1350,[1] have been more or less thrown open by the Judges of the Probate Court; but the accommodation for searchers was long inadequate, and the restrictions are harassing. But, indeed, for the personal courtesy of Mr. Smith, the able and most obliging antiquary who presides over the Literary Search Room at Somerset House, matters would indeed go badly with any enquirer.[2]

The Wills enrolled in the Court of Hustings of the "City" of London begin much earlier than any other series (1258), and an excellent calendar of them is being issued by the Corporation.

There are many printed collections of wills, all of which will be more or less useful to the beginner, as accustoming him to the phraseology of this class of documents.

Among the best known are :—

(1) *Testamenta Vetusta*, by (Sir) Harris Nicolas, London, 2 vols. 8vo., 1826, which however deals almost exclusively with well-known and well-born families.

(2) *Wills from Doctors' Commons*, by J. G. Nichols

(3) *Bury (Suffolk) Wills and Inventories*, by S. Tymms, published by the Camden Society in 1850.

(4) *Testamenta Eboracensia*, published by the Surtees Society.

(5) *Wills and Inventories illustrative of the History of the Northern Counties*, published by the Surtees Society.

(6) *Lancashire and Cheshire Wills*, 3 vols. (Chetham Society), by the Rev. G. J. Piccope, 1857-61.

---

[1] For a report on these, see 1857 Report, p. 237.

[2] As to searching here see *post*, p. 121.

(7) *Early Norfolk Wills from the Norwich Registry*, by John L'Estrange (being all the interesting wills from Regr. Heydon), 1370-1383. *Norfolk Antiquarian Miscellany*, i. pp. 345-412.

To these may be added:—

(8) Griffiths' *Index to Wills proved in the Court of the Chancellor of Oxford*, 8vo., 1862.

(9) *Wills of the Pastons and their Connections*, published by Mr. Gairdner in his edition of the *Paston Letters*.

(10) *Calendar of Lambeth Wills* (1313-1644): *Geneal.* (1st series) v. p. 211. This is to the *Vacancy Wills*, and is specially valuable, because it gives references to all wills proved during the vacancy of any See belonging to the Province of Canterbury. See also *Catalogue of Lambeth Administrations: Geneal.* (N. S.) i. p. 80.

(11) *Calendar of Early Suffolk Wills*, Ipswich Registry, 1444-1620. *East Anglian*, (N. S.) begun vol. i. p. 340.

(12) *Some Wills in the Public Record Office. Geneal.*, iii. (N. S.) pp. 122-6, 185-7.

(13) *A Catalogue of the Yorkshire Wills at Somerset House*, 1649-1660. Yorkshire Archl. and Top. Association.

(14) *Index to Bundles of untranscribed Wills at York*, temp. Charles I. Id., not yet issued.

(15) *Calendar of Northamptonshire and Rutland Wills* (1510-1652), now being published by Mr. Phillimore in his *Index Library*; as to which see Index to this work.

(16) *Abstracts of Somersetshire Wills*, by the late Rev. F. Brown, F.S.A., to be printed at £1. 1s. by Mr. F. A. Crisp, Grove Park, Denmark Hill, S.E.

(17) *Testamenta Lambethana, &c.*, (1312-1636) by Dr. Ducarel, folio, pp. 138. Sir T. Phillipps, Middle Hill Press.

(18) *Early Lincoln Wills* (1280-1547). Williamson, 290, High Street, Lincoln.

Of course there are immense MS. collections from wills in the British Museum and elsewhere. For a detailed list of these I will refer my readers to Sims' *Manual*, pp. 344 et seq. He has, however, omitted perhaps the finest, viz., D'Ewes' *Collection of Norfolk Wills*, which is Harleian MS. 10, and which contains many thousand notes of the highest value to East Anglian antiquaries. John L'Estrange, to whose first vol. I have referred above, also made very large collections of Norfolk wills (four vols. folio), which are now in my possession, and are being indexed.

Following the wills, the *Inventories and Accounts* rendered by the executors in pursuance of their oaths on proving the wills, are often of the highest interest, as giving the minutest details of the furniture, plate, and other effects of the deceased. The accounts of Richard, Bishop of London, in 1303 (published by Archdeacon Hale for the Cam. Soc.), and of Sir John Fastolf (printed by Mr. Gairdner in his *Paston Letters*), are both very fine specimens of this class of record.

CHAPTER XII.

Manorial Records,[1] Court Rolls, &c.

I HAVE no hesitation in saying that where a long series of Court Rolls of a manor have been preserved, they are the most valuable, interesting, and easily consultable class of records the topographer can have. The genealogist, and especially he who interests himself in the descent of those who do not belong to armigerous families, will often find them the only way of tracing pedigrees before the Church Register begins.

But, alas, not one manor in twenty has a good series of the rolls, and of these not one in twenty is accessible to the enquirer. Many stewards have the mistaken idea that the old rolls are nuisances, as containing entries which may be used against their lords some day on questions of commons or customs,

[1] The best book to give one an idea of the management of manors in the old days is the *Domesday of St. Paul's*, published by the late Archdeacon Hall for the Camden Society, 1858.

and very rarely allow any one to consult them without the special consent of the lord.

If this consent *is* obtained, the office of a busy solicitor (who will of course expect to be paid for the production) is not the place in which an antiquary loves to linger, at so much an hour. When, however, the rare opportunity occurs, the topographer should never miss it, but, setting all other business aside, copy, copy, and copy, till he has as much as he possibly can.

The rolls contain two distinct classes of information. The Lete Roll (usually part of the general roll) is practically the presentments of the offences of the tenants, whether for assault and battery, blood drawing, theft, &c., or for breaches of sanitary and other wise regulations. The general presentments are of the deaths of the tenants and the consequent admissions of their heirs or legatees (in which case the will is enrolled), of sales and mortgages, and other regular conveyancing business. For specimens of the "Extents" of a manor see 1800 Report, pp. 145-6.

Several vols. of manor rolls have been printed, *e.g.*—

(1) *The Court Leet Records of the Manor of Manchester in the Fifteenth Century,* by John Harland, Chetham Society, 2 vols., 1864-5, which as its name implies, relates only to the Letes, as also does—

(2) *Extracts from Court Rolls of the Manor of Wimbledon,* extending from 1 Edward IV. to 1864, published by the Wimbledon Common Preservation Committee, London, 1866, which is specially valuable to those interested in getting up evidence for the preservation of open spaces. The "remainder" of the work is now in the market very cheap, and is well worth buying, for it gives the Latin on one side, and the translation the other.

(3) *The Court Rolls of Cressingham* (Norfolk.) Privately printed by H. W. Chandler (Oxford), 1885.

(4) *The Court Rolls of Socton with the Soke* (Suffolk.) *East Anglian,* vol. iv. p. 3 et seq.

There are many Court Rolls in the Augmentation Office, indexed on Shelf 5 of Division J of the Round Search Room.[1]

---

[1] And see 1837 Report, p. 186 a; 20th Report, p. 80.

So there are in the Bodleian, the British Museum (especially the Additional MSS.), and elsewhere.

A translation of the "compotus" of the manor of Newton, near Wisbeach, for 1395, is printed in *East Anglian*, vol. iv. p. 69 et seq. There is much about the enfranchisement of bondsmen in blood or villeins among the Exchequer Commissions and Depositions, q. v., p. 47.

CHAPTER XIII.

Grants from the Crown, Privileges, Titles, &c.

SPEAKING roughly there are four classes of records which give information on the subject matter of this chapter;—(1) The Charter Rolls; (2) the Originalia Rolls; (3) the Close Rolls; and (4) the Patent Rolls.

The *Charter Rolls* begin with 1 John (1199), and end with 12 James I., but really much earlier, there being few entries after Richard II.

They contain grants of fairs, markets, and free warren privileges, incorporation charters, &c., &c., but their nature can best be judged by inspecting the vol. for King John (1199-1216) which was printed in 1837, in folio, by the Record Commission.

There is also a printed calendar, which was published in 1803 (sells for 15s.), and extends from 1 John (1199) to Edward IV. (1483), but it is not a good one. It includes also a Calendar of the *Inquisitions ad quod damnum* from 1 Edward II. to 38 Henry VI. These inquisitions were often the precursors of King's charters, licenses in mortmain, &c., &c., the King frequently directing an inquisition or enquiry to be made, whether it would be an injury to himself or his subjects if he made a certain grant or gave a certain license.

On this subject, of course, the Records of the Treasury of the Exchequer are most valuable.

Whenever any service, rent, or salary was reserved by any grant or charter, special memoranda of it were entered in the *Originalia Rolls* by the Exchequer officials, to whom they were transmitted by the Chancery.

They are in fact rolls which contain all the information that the officers of the Exchequer needed for the carrying on their duties of collecting the King's revenue, and from them may be obtained :—

(1) The names of nearly all officials, and certainly of all who under any circumstances might have to account for anything to the King.

(2) Enrolments of all grants by charter or patent under which were received anything to be paid to the King.

(3) Letters of denization and some pardons,[1] especially when a man had pardon for his life, but not for his lands.

The Originalia from 20 Henry III. to 51 Edward III. (1235-1377) were abstracted for the Government, and the abstracts printed in two vols. fo., in 1805-10, a work which now sells for 12s.

They have been partly transcribed for the reign of James I.; see 1819 Report, pp. 41, 42, 48. For a facsimile of part of them see same Report, No. xxiii.

A continuation of these abstracts from 1 Richard II. to the end of James I., now forms Additional MSS. 6363, 6387, at the British Museum. I do not know if there is a duplicate of this Calendar at the Record Office.

The first vol. of a printed work, Jones'[2] *Index to the Originalia and Memoranda*[3] *of the Lord Treasurer Remembrancer's side of the Exchequer* (2 vols. fo., 1793) gives many entries for the period between Henry VIII. and Anne (inclusive), and is a

[1] Pardon Rolls exist from 24 Edward I.; see 1837 Report, p. 70 a; and for Richard III. to James I. (unindexed), id. p. iii.

[2] Apparently they were not by Jones himself, but by a Mr. Chapman, who left them to him. *Cooper*, i. p. 345.

[3] For an account of the Memoranda Rolls see 1837 Report, p. 177.

very useful work, though the index locorum is very poor. A copy is now worth about 9s.

The second vol. gives reference to various Exchequer entries, such as enrolments of charters, grants, and patents to several religious houses and to cities, boroughs, towns, companies, colleges, and other public institutions, from the earliest[1] period, &c.

The Records of the *Treasury*[2] are of course very voluminous. A report on them will be found in the 7th Report, Appendix II. Other references likely to be useful are—Extracts from the Minute Books of 1634-5 to 1717; id. p. 65; Inventory of Treasury Warrants, 1548-1834, 8th Report, p. 189; List of Calendars, 24th Report, pp. 73-80; List of Lord High Treasurers, 25th Report, pp. 61-70. There has also been separately published (1871, imp. 8vo.) a Calendar of Treasury Papers.

The *Close Rolls*,[3] so called because they were registers of letters closed or sealed up, and directed to individuals, in contradistinction to the Patent Rolls, which recorded those "patent" or open for everyone to read and take notice,—contain a vast number of entries of the highest interest, such as proclamations suppressing riots and tumults, orders to the sheriffs on all sorts of matters, directions for raising and collecting subsidies.

The bulk of these rolls is immense. Mr. Selby in the Introduction to his *Lancashire and Cheshire Records*, p. 9, says there are 18,342 rolls from 1204-1879, which would fill if printed 450 folio vols. of 1,000 pp. each, a fact which he rightly commends to the consideration of those who talk about printing the records *in extenso*.

The roll from 6 John (1204) to 11 Henry III. (1227) was printed in two vols. folio by Duffus Hardy for the Government,

[1] Not between 1509 and 1714.

[2] The most comprehensive account of the early documents of the Exchequer and Treasury is to be found in *The Ancient Exchequer of England, the Treasury, &c.*, by F. S. Thomas, London, 8vo. 1848.

[3] For a specimen calendar for 12 Henry III., see 26th Report, p. 48, and for note how made up, B, Y, E, or N, see 2nd Report, p. 38.

H

under the title of *Rotuli Lit. Claus*, and now sells for about £1. 10s. The rest of the reign of Henry III. has been excellently calendared by the late Mr. Sharp, in ten vols. folio, the MS. of which is in the Round Room of the Record Office.

Hardy published a description of these rolls in 1833, which was only a reproduction of the Introduction to his published vol. i. A facsimile of an entry on this roll is in the 1819 Report, No. ix.

All the Parliamentary Writs and Summonses to Parliament are entered on the dorses of these rolls, which also contain many private deeds enrolled.

There is said to be an index to these rolls in the library of Lincoln's Inn, and the 1800 Report, p. 92 b, gives a list of all the abstracts and indexes up to that date. As to the later indexes see the list of Record Office Calendars in the 2nd Appendix to the 41st Report.

An official index from Elizabeth to the present time (?) was kept at the Enrolment Office.

The *Patent Rolls*[1] extend from 3 John (1201), and contain innumerable grants of offices and lands, fairs and markets, confirmations, licenses to crenellate or fortify, licenses for the election of bishops,[2] abbots, &c., creations of peers, pensions,[3] &c., and of later years the patents for inventions.[4]

On the backs are noted the Commissions to the Justices of

---

[1] For a facsimile of an entry on this roll see 1819 Report, No. viii., and at page 73 of the same Report there is a specimen of a short tabular calendar of these rolls. In the 1800 Report, pp. 90-1, there is a list of all the abstracts, and indexes of them. Specimens of proposed calendars of these rolls will be found in the 2nd Report, p. 273, and the 26th Report, p. 66, (1 Henry III.)

[2] For appointments of Bishops enrolled on the roll, temp. Charles II., see 45th Report, Appendix I.

[3] For a calendar of volumes containing offices, pensions, &c., Henry VIII. to Charles II. see 2nd Report, p. 205.

[4] For a calendar of the Specifications see 6th Report, p. 116 ; 7th Report, p. 101 ; and 8th Report, p. 82. For those from 8 Anne see 1837 Report, p. 118 b. There are "Specification Rolls" from the Restoration, with complete indexes ; see 1837 Report, p. 113.

the Peace, &c., which pass the Great Seal, commissions to try causes as to land. &c. The best description of the Patent Rolls is that by T. D. Hardy, which he printed privately in 1835.

A folio volume of so-called calendar was published by Government in 1802, but obviously was of the scantiest selections only, the rough note book, indeed, of some office clerk in the reign of James I. (*Cooper*, i. p. 297). Hardy in his account of the Patent Rolls says that this calendar does not give a tenth part of the entries of John and Henry III., while Cooper (i. p. 299) says not more than a fourteenth. It now sells at about 5s.

In 1835 a folio vol. of all the Letters Patent extracted (by Sir T. Duffus Hardy) from the Patent Roll, but not forming all its contents, covering the period 1201-1216, was printed by Government, with admirable indexes.

At the Record Office there are three vols. of full modern Calendars of the Patent Rolls for Henry III., but not indexed.

Very good lexicographical calendars of the Patent Rolls, from 1 to 6 Edward I., are printed in the 42nd to 47th Reports.

Except for the little help obtainable from Palmer's Indexes, mentioned below, there is nothing to bridge over the awful gap between Henry III. and Edward V., at which latter date an excellent new calendar (with, however, a most mysterious and puzzling index) was printed in the 9th Report, pp. 148 et seq. This covers two reigns, Edward V. and Richard III., and is continued in MS. in the Round Room of the Record Office; for Henry VII. (11 vols.), Henry VIII. (14 vols.), and part of Edward VI.[1]

---

[1] Unluckily, however, these vols. are not indexed, a fact which terribly discounts their value. If the authorities at the Record Office would only realize the fact that one indexed calendar is worth two unindexed, and that lexicographical indexes could be contracted for out of the Office at less than 10s. per 1000 references, some improvement would soon take place. Again, the puerility of making alphabetical indexes only out of cut-out slips is most vexatious. A very slight extra expenditure of labour would save an immensity of searchers' time.

From Henry VIII. there are official MS. Indexes Nominum to the Patent Rolls, and there are more or less complete (though unofficial) Indexes Locorum, known as *Palmer's Indexes*,[1] to be found in the Long Room. At pp. 205-209 of the 2nd Report is a calendar of the enrolments of all Letters Patent, granting offices, pensions, and honours, which are easier to search. The "Signet" Indexes, which give clues to the Patent Rolls, are being printed from 1584-1624, by Mr. Phillimore in his *Index Library*, as to which see the Index to this book.

Besides these four sets of rolls you should, if you are looking for early matters, refer to the *Pardon Rolls*, which exist from 24 Edward I., as to which see 1837 Report, pp. 70 a and 111.

The *Hundred Rolls*, another record of the Exchequer, consisting of inquisitions taken under a commission, dated 1274, may be broadly stated to be the results of inquisitions into all rights of manor, warren, chase, fishery, toll, market, &c., claimed at the date of the commission, which was issued to put an end to various extortions and tyrannies which had then sprung up.

These rolls have been printed in extenso, viz., in 1818, and are scarce, costing £4 or so. See 1st Report, p. 145.

A sequel to these Hundred Rolls were the *Placita de Quo Warranto*, temp. Edward I., II., and III.; which were the trials ordered to test the justice of the claims mentioned in the Hundred Rolls. For a facsimile of part of one see 1819 Report, lxii. It was printed in 1818 in a thick folio volume, and is scarce, selling for £2. 10s.

Then there are the *Inquisitiones ad quod damnum*, already described at p. 95.

If you are specially looking for grants of offices from the Crown, you should, besides the four sets, go through the Pell

---

[1] For a description of *Palmer's Indexes* see Selby's *Lancashire and Cheshire Records*, p. 463, and the 20th Report, pp. 184-8.

Office Issue Rolls, the Patent Books (the Auditors' series and the Pell series separately),[1] the Privy Seal Books[2] (similarly divided into series), the Docket Books of the Privy Signet Office;[3] and the indexes of the same office, both of which begin 1584, should also be consulted. It is to be remembered however that these Signet Indexes refer to the Signet Bills (Home Office).

The Lord Chamberlain's Records should also be searched. They are at the Record Office, but are not yet open to the public without a special permit. The Privy Council Records are still at Whitehall, and can only be seen by special permission, which is obtainable by application to the Lord Chamberlain; but as the searcher must have a special introduction, and the officials have apparently the greatest and most absurd jealousy of allowing their records (even of the reign of Elizabeth) to be seen, the effort is hardly likely to be successful.

Of course these records of the Privy Council ought long ago to have been brought to the Record Office, being Domestic State Papers of the highest interest.

As to *Crown Lands*[4] the calendar of the Decrees of the Court of General Surveyors, 34-38 Henry VIII., pp. 166-196 of the 30th Report will be found useful. See also 1800 Report,

---

[1] For printed calendars see 2nd Report, p. 195; 4th Report, p. 165; 5th Report, p. 245; and 6th Report, p. 227. For an inventory of the books of the Pell Office see 7th Report, p. 213.

[2] For enrolments of Privy Seal letters, 1594-1741, see 2nd Report, p. 211, and for Privy Seals during the Commonwealth, see 5th Report, p. 245; from 1634-1711, see 30th Report, pp. 360-583; and from 1625-1632, see 43rd Report, pp. 1-205; from 33 Henry VIII. to 22 James I., see 1837 Report, p. 15 a.

[3] For a list of the records of the Signet Office see 24th Report, p. 86; see also p. 100, as to some of them being printed by Mr. Phillimore. Cambridge University MSS. D. d. iii. 53, is a volume of precedents for this office.

[4] An early roll as to Crown Lands committed to farm in 10 Edward I. is referred to in 1819 Report, p. 187.

p. 163. The entry book of Crown Leases, from 1599-1696 is calendared in the 2nd Report, pp. 209-211. A calendar of Grants and Leases, temp. Charles I., in seven large bundles, with indexes, is mentioned in 1819 Report, p. 36.

Grants of Leases and Crown Lands of a recent date are set out in a rather scarce book, printed in 1788, called *An Account of all Manors, Messuages, &c., held by lease from the Crown* . . . . . . *also a Calendar of the Surveys of the Estates of Charles I., &c.*, arranged in counties in a tabular form, but this is very scarce.

There is much in the Record Office relating to *Royal Forests*. A calendar of the rolls, &c., relating to them, is at pp. 126-7 of the 20th Report, and there is a descriptive catalogue from Henry III. to James I. at the Record Office. The Perambulation Roll for 7 Edward I. is mentioned at p. 54 of the 2nd Report, and one for Windsor (28 Edward I.) id. p. 64.

For peerage and baronetage claims, the searcher should, besides Dugdale, Burke, Collins, Nicolas, and Foster, the standard works on the subject, refer to *Parliamentary Writs and Writs of Military Service*, by Sir F. Palgrave, two vols. folio, 1834, which has magnificent indexes. It now sells for about £3. 3s.

The grants of peerages and baronetages are entered on the Patent Roll. There is a calendar of the creations of peers from Richard III. in the 47th Report, p. 78; and of baronets from 1611, id. p. 125.

CHAPTER XIV.

The Record Office, British Museum, Probate Registry, City of London Records, Lambeth Library, Heralds' College, the Bodleian and other Libraries.

THE RECORD OFFICE.

FEW of those who frequent the comfortable and convenient Literary Search-room at the Public Record Office in Fetter Lane, have any idea of the inconveniences which attached to those who, like myself, used to search twenty-five years ago in a long unpleasant room, with low tables and high backless forms, which cramped the searcher's legs if he were anything above a dwarf in stature. A searcher then had to pay a shilling for each search, and had to make his copies in pencil: a barbarism to which, however, the authorities have compelled us to temporarily revert recently; and though the bad old days of extortion were over, those of civility and help had not yet come.

Now-a-days, it would be difficult to find a place where study and search can be carried on more easily and pleasantly than at the Public Record Office, and certainly at no place are the officials, from the highest to the lowest, more courteous or more willing to help. The beginner, stumbling along, and only half conscious as to what he is looking for, is as well treated and listened to as patiently as the *habitué*; and with the single exception of lunatics, who want information about unclaimed millions "in Chancery," who are sternly kept at bay by a standing notice in the lobby, all searchers, however different their objects, are made welcome.

The cause of all this is easily explained. There is hardly a clerk in the office who does not take an active and intelligent interest in some class of records, and who has not contributed something [1] to their history. In fact, the present and past staff may be said to be teachers of a record school or university, to the very great advantage of the amateur students.

It is very pleasant to consider how different the present state of things is to that which is disclosed by the amusing controversial tracts, from 1831-1840, when Cooper, Cole, Palgrave, and Hardy were at one another's throats on various issues, when the machine worked with as little harmony as it possibly could without stopping, and when, it is obvious, jobbery and incompetence were rife. I have a good collection of these tracts, and often amuse myself by reading them, and trying to get to the rights of the attacks and counter attacks they contain, but ultimately come to the conclusion that one thing only is certain, and that is that speaking generally—of course there are exceptions—the waste of public money was only equalled by the badness of the work for which it was paid. Useful hints and stray facts may sometimes be picked up among the *scoriæ* of this literary dust bin, but I doubt if it is worth while raking it up—certainly not here!

Now-a-days, the office is kept up most efficiently and very economically; and if I venture further on upon a few criticisms and minor grumbles, I trust they will be taken in good part, the more especially as the present Deputy Keeper has only just taken office, and, except in one instance, is not responsible for anything I mention.

Full particulars of the rules and regulations of the office are given at p. 109, and the plans on pp. 112 and 114 will enable a searcher to put his hand on any of the more important calendars, and so begin his search. All the numbered calendars and indexes in the Record Office are referred to direct in the Index to this work, the number after each being the reference to the volume, and if this is given to the attendant, he will at once bring it to the searcher. It will be as well, however, before

[1] I need hardly remind my readers of such works as Mr. J. Gairdner's *Paston Letters*, Mr. J. J. Cartwright's *Wentworth Papers*, Mr. Luke O. Pike's *History of Crime in England*, Mr. Hubert Hall's *History of the Customs* and *Society in the Reign of Elizabeth*, Mr. Ewald's *Handbook to the Records*, and last, but far from least, Mr. Walford D. Selby's most valuable works so frequently cited in this book.

beginning any lengthened search, for the beginner to study at home all or any of the following works:—

The Repertorie of Records, by Tho. Powell, 1622, 1631. [Said to be compiled from Agarde's Papers.]

An Index to the Records, 1739. [This is the first attempt at a digested index under subjects.]

An Account of the most important Public Records of Great Britain, by C. P. Cooper, 2 vols. 8vo., 1832. [Very useful, but very discursive, and practically without index.]

The Ancient Exchequer of England, &c., by F. S. Thomas, 8vo., 1848.

Official Handbook to the Public Records, by F. S. Thomas, Secretary of the P. R. O., 8vo., 1853. [A store-house of information, but vilely arranged and worse indexed. Take, for example, p. 214— not *one* of the sub-items on it is in the index. I hear a new edition is in progress, and only hope it may have a real index.]

A Manual for the Genealogist, Topographer, &c., by R. Sims of the British Museum, 8vo., 2nd edition, 1861. [This is much the best that has been yet issued, but the index might be improved.]

Our Public Records, by A. C. Ewald, F.S.A., 8vo., 1873. [Useful for references to recent Record reports, and a good glossary.]

Notes of Materials for the History of Public Departments, edited by F. S. Thomas, Esq., London, 1846, 1 vol, small folio. [Contains useful references to State Papers.]

Record Commission, Proceedings of, from June, 1831, to August, 1833, 1 vol. folio. [A complete set is very scarce, only very few perfect copies having been preserved. To a collector this is the "great auk's egg" of record publications.]

Papers and Documents relating to evidence taken before Select Committee of House of Commons, 1837, 1 vol. 8vo.

Court of Wards and Liveries, specimens of the books containing the Sales of Wards and Marriages, by Sir Francis Palgrave, London, 1834, 1 vol. 8vo.

Lancashire and Cheshire Records, vols. 7 and 8 of the Record Society's Publications, by W. D. Selby, 8vo. 1882. [Contains alphabetical class lists of the Records of the Duchy of Lancaster, Palatinate of Lancaster, and the County Palatine of Chester.] See also his Norfolk Records.

On beginning his search the novice will find himself greatly hampered by the recent rule that he must not use ink—not even from a stylograph. This is a barbaric rule, which I hope will soon be altered. It, of course, behoves all searchers to take the greatest possible care of the documents entrusted to them, and all officials to see they do so; and no reasonable man would object to very severe rules, such as that any person inking a record, or even the desk on which it rests, should be suspended for the first offence and debarred from ink again for the rest of his life for the second. But considering how extreme is the inconvenience to all workers of having to take pencil notes first, and then having to transcribe them at home into ink, thus practically doubling the chance of error as well as the time and work necessary for any search, it does really seem very hard that the careful now have to suffer for the past sins of the careless. If the paid transcribers are still allowed to copy for the public in ink, why should not the public be allowed to copy for themselves in the same material? A rule like that suggested above would answer every purpose, especially if supplemented by a regulation that stylograph pens (*not* to be filled on the premises) only should be used. As it is, I am confident that much useful research will be stopped,[1] and for no practical purpose. The authorities at the British Museum do not insist on pencil only being used, and the MSS. there are, as a whole, of greater individual value than any single membrane—say of a De Banco Roll.

Another thing he will find worrying and troublesome is the question of ventilation, which makes the Round Room a veritable rheumatism trap in the winter, though the new Deputy Keeper has already worked wonders to amend this.

Nor will he like the way in which many of the new "Indexes" have been compiled under alphabets only. When cut-up slips are used, it is scarcely more trouble to sort them lexicographically than to sort them alphabetically, and the

[1] Luckily for me, a long Calendar I had on hand was almost done before the rule came into existence; for if it had to be done in pencil first, it certainly never could have been done, and I think this is a fair sample of work that will now be stopped.

scarcher's extra time saved is immense. Indexing is a thing that might be well contracted for out of the office, for it is like breaking a butterfly on a wheel to employ skilled labour (which might well be applied elsewhere) in making indexes, which could be done, and *well* done (credo experto), by contract, at from 5*s.* to 7*s.* 6*d.* per 1000 references.

But it will be the *want* of indexes which will bother him most. Take, for example, the Feet of Fines—the chief records of the transfer of landed estates, and as invaluable to the topographer as to the genealogist. With the few exceptions mentioned at pp. 40-2, these documents are literally unindexed to the reign of Henry VII., and practically unindexed from that date. It is true there are so-called " indexes " on the shelves of the Round Room, but they are nothing more than lists of the names of the parties arranged under counties for each term. The labour, therefore, of a general search in any one name, even for one county only, is simply prohibitive, and the books are seldom searched except for likely periods. Even then, the books are very old and their leaves very tender and worn, and each search, during which nearly every page of each volume is touched, helps to hasten its end; and when they are gone, even this help will be gone too, and the files themselves will have to be searched, and incalculable further labour created. A plain modern transcript of these " indexes," which could be indexed volume by volume, would not be so very expensive ; and if printed in small type could be made to pay its own cost of printing.

As many of my readers, and especially such as live in the country, may never have the chance of making a personal search at the Record Office, it may be as well for me to give the names and addresses of some of the best-known Record Agents,[1] who will, in nearly every case, undertake searches at the British Museum and other London repertories. Those to whose names a † is prefixed have published antiquarian works.

[1] It may save me trouble to state here that, though I happen to be a solicitor, I am *not* a record agent, and do not undertake record work either in or out of my own county.

Record Agents.[1]

Boyd, W., 12, Sloane Terrace, S.W.

Fox, Miss Rita, 1, Capel Terrace, Forest Gate.

†Greenstreet, J., 16, Glenwood Road, Catford, S.E.

Grigson, Mrs. F., 45, Alma Square, St. John's Wood.

†Hart, W. H., Hammersmith.

Hardy and Page, Messrs., 22, Old Buildings.

Heintz, A. F., 11, Great Western Road, W.

†Hewlett, M. H. and W. O., 2, Raymond's Buildings (solicitor).

Kirk, R. E. G., 27, Chancery Lane.

†Phillimore, W. P. H., 124, Chancery Lane (solicitor).

†Smith, Miss L. Toulmin, 101, Southwood Lane, Highgate, N.

Vincent, J. A. C., 17, Hart Street, Bloomsbury, W.C.

Walford, Miss, 7, Hyde Park Mansions, Edgware Road, W.

Transcribers.

Collier, Miss, 83, Charterhouse Street.

Hopper, Miss, 9, Cato Road, Brixton.

Indexer.[2]

Macfarlane, Miss, care of Mr. J. D. Macfarlane, 12, Catherine Street, W.C.

The charges of record agents vary very much indeed. Experienced agents can command a very much higher price than younger men, for the simple reason that their experience enables them to go straight to the class of records likely to afford the information wanted by their clients, and their services, therefore, are often really much cheaper in the long run than those of beginners who nominally work at a much lower rate, but who take very much more time to obtain the same result : the client in effect paying for their growing experience.

For plain copies of documents, to which the client himself supplies the reference, he need not pay more than 6d. per folio of seventy-two words of writing before 1600, and 4d. per folio for later documents.

---

[1] It may be of use to some to note that the chief Specialists at the Bar, for antiquarian subjects, are:—C. Elton, Q.C., F. H. Jeune, Q.C. (Ecclesiastical law), J. W. B. Willis-Bund (fishery rights), Stuart-Moore (foreshore and manorial rights).

[2] Indexing charges vary upwards, from 7s. 6d. per 1000 references to names or places.

The charts on pp. 112 and 114 show the position of most of the important Calendars and Indexes in both Search-rooms; the letters referring to the lists opposite each chart.

The following are the regulations by which the Record Office is governed :—

1. In these rules and regulations, the word Record shall include a set of records or documents comprised in a single reference in the official calendars, catalogues, or indexes.

2. The Search-rooms shall be open to persons desiring to inspect or search records or documents on every day, except Sunday, Christmas Day to New Year's Day inclusive, Good Friday, Easter Eve, Easter Monday, Easter Tuesday, Monday and Tuesday in Whitsuntide, Her Majesty's Birthday, Her Majesty's Coronation Day, and days appointed for public fasts or thanksgiving.

3. The hours of admission and attendance shall be from 10 to 4 o'clock, except on Saturdays, when they shall be from 10 to 2.[1]

4. Every person proposing to inspect or search records or documents, shall write his or her name and address, daily, in the attendance-book kept by the hall-porter.

5. No umbrellas, sticks, or bags, shall be taken into any Search-room, and no parcels shall be placed upon the tables.

6. A separate ticket shall be clearly written and signed by every person desiring to inspect or search any record or document for each record or document required, and such ticket shall be given by such person to the officer in charge of the room before any record or document can be produced to the applicant.

7. No person shall have more than three records or documents, inclusive, out at a time.

8. Records and documents when done with, shall forthwith be returned by the person to whom they have been produced, to the officer in charge of the room, or to one of the attendants, in exchange for the tickets referring to them, and every such person shall be held responsible for the records or documents issued to him or her, so long as his or her ticket shall remain with an officer of the Public Record Office.

9. A fresh ticket, clearly written and signed by the person to whom records or documents have been produced, and bearing the words " kept out," shall be required for every record and document kept out from one day to another [2] for the convenience of such person.

---

[1] The early closing on Saturday is very hard on those who are engaged all the week, and whose only spare time is the Saturday half-holiday.

[2] Under special circumstances it might well be arranged that documents should be left out for seven days to known searchers, especially to those who are doing amateur calendaring.

10. No person shall lean upon any records, documents, or books belonging to the Public Record Office, or place upon them the paper on which he or she is writing.

11. No person other than an officer of the Record Office shall make any mark, in pencil or otherwise, upon any record, document, or book belonging to the Public Record Office.

12. Ink shall not be used in the Public Search-rooms by any person admitted thereto for the purpose of inspecting or searching records or documents.

13. Tracings of records or documents shall not be made by any person without specific permission from the officer in charge of the room.

14. A list of all calendars, catalogues, and indexes intended for the use of the public shall be kept in each of the Search-rooms, and shall be revised from time to time. Calendars, catalogues, and indexes, not mentioned therein, or withdrawn therefrom, shall not be produced in any Search-room without an order from the Deputy Keeper.[1]

15. Records not mentioned in any such list, and records in course of arrangement, shall not be produced without an order from the Deputy Keeper.

16. Records and documents of exceptional value, and records and documents in fragile condition, shall be produced singly, or subject to such conditions as the officer in charge of the room shall, in the particular case, think requisite for their safety and integrity.

17. Documents deposited in the Public Record Office by other Departments shall be produced only on the order of their respective heads, or subject to such conditions as they, with the concurrence of the Master of the Rolls, shall from time to time impose.

18. Persons admitted to the Search-rooms shall replace the calendars, catalogues, and indexes used by them on the shelves as soon as they are done with.

19. Records, documents, books, or other articles belonging to the Public Record Office, shall not be removed from one room to another without the specific permission of the officer in charge of the room.

20. Silence shall be maintained in the Search-room, as far as possible.[2]

21. The officer in charge of any Search-room shall be empowered to exclude persons from the Public Record Office for any of the

---

[1] The latter part of this rule is simply stupid.

[2] It could be wished that this rule could be better enforced. If any two seachers want to carry on an animated conversation in high voices on literary or domestic subjects, they would find the corridor admirably adapted for that purpose. The gentleman who reads records to himself in a loud voice with chuckling accompaniments expressive of high delight or disgust might be cautioned, and disqualified after two cautions.

following reasons :—Wilful breach of any of the foregoing rules and regulations, persistent disregard of the officer's authority, damage of any sort to any record or article belonging to the Public Record Office; conduct, language, habits, unseemly dress, or any other matter offensive, or likely to be reasonably offensive, to others using the Public Record Office. Provided always, that the exclusion of any person shall be forthwith notified in writing with the cause thereof to the Deputy Keeper, who shall inquire into the circumstances, and whose order, unless reversed by the Master of the Rolls, shall be final.

22. Inasmuch as a large number of Records of the Courts of Law and Equity have been of late years transferred into the Public Record Office, and are there frequently consulted for other than literary and historical purposes, fees shall be paid according to the table hereunder set forth for the production of such records, unless the applicant satisfies the Deputy Keeper that such production is required solely for the purpose of literary or historical research.

The other fees specified in the said table shall be payable as therein mentioned.

TABLE OF FEES, payable in every case by Stamps.

	£	s.	d.
For inspecting a record of any Court of Law or Equity, of a later date than the year 1760. per diem - - -	0	1	0
For a general search, to include the inspection of all the records in any one class referring to a particular suit, action, or matter, of a later date than the year 1760, per diem - - - - - - -	0	2	6
For authenticated copies of records or documents to the end of the reign of George II., per folio of 72 words -	0	1	0
For authenticated copies of records or documents after the reign of George II , per folio of 72 words - -	0	0	6
For authenticated copies of plans, drawings, &c., per hour	0	2	6
For attendance at either House of Parliament or elsewhere, to produce records for the purpose of evidence, per diem . - - - - - -	2	2	0
For attendance on the Master of the Rolls on a Vacatur -	0	10	0
For attendance to receive mortgage money - - -	0	5	0
For attendance on payment of mortgage money - -	0	10	6

These rules and regulations shall take effect on the 1st of May, 1887, in substitution for the existing rules and regulations, which shall as from that day cease to have any operation.

(*Signed*)   ESHER, M. R.

A.—Chart of Round Room, Record Office.

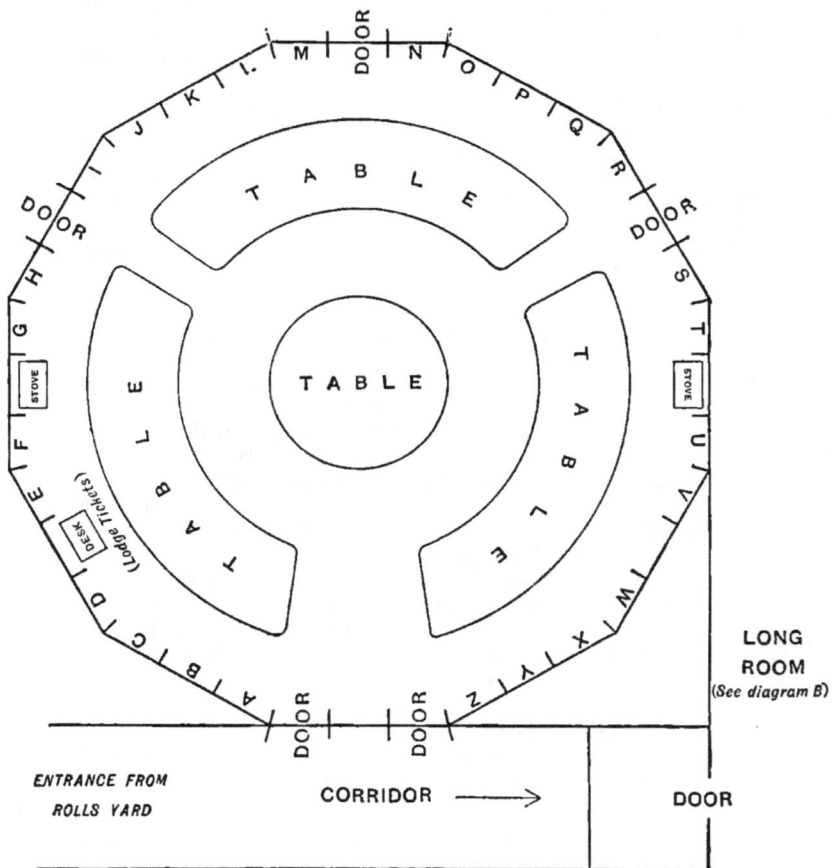

ROUND OR LITERARY SEARCH ROOM.—REFERENCES TO CHART
OPPOSITE.

WALFORD D. SELBY, *Superintendent.*

A 1, 2, 3, 4 Folio Publications of the Record Commission (Domesday, Hundred Rolls, Statutes, Rolls of Parliament, Inquisitions P.M., &c.)

B 1, 2 . Acts of Parliament of Scotland.

3, 4 . . Indexes, Calendars, &c., of Court of Requests. Placita Forestæ, Star Chamber, &c.

C 1 . . Various books of Reference, Archæologia, &c.

2, 3 . . Dugdale's Monasticon and Reports of Deputy Keeper of the Records.

4, 5, 6 . Agenda Books (Exch. Q.R.), Signet Bills (Chapter House).

D 1—6 . Master of the Rolls series of Publications, (" Chronicles and Memorials").

E 1—5 . Ditto.

F 1, 2, 3 . Exchequer Rolls of Scotland, Register of Privy Council of Scotland, &c.

4, 5, 6 . Indexes to Warrants of Attorney and Cognovits.

G 1—11 . Exchequer Indexes, Subsidies, Queen's Remembrancer's Miscellanea, Ministers' and Receivers' Accounts, Escheators' Accounts and Inquisitions, &c.

H 1—7 . Calendars to State Papers, Foreign and Domestic, &c.

I 1, 2 . . Ditto.

3, 4, 5 . Various Books of Reference, Dictionaries, &c., Pipe Roll Society's Publications, and Octavo Publications of the Record Commission.

J 1, 2, 3 . Catalogues of Enrolments in Exchequer of Pleas.

4, 5 . . Liber Dockett, Exchequer of Pleas.

K 1, 2 . Ditto.

3, 4 . Agarde's Indexes.

5 . . Indexes to Feet of Fines, various counties.

L 1 . Calendars to Placita de Banco, &c.

2, 3 . Catalogue of Pleas of the Crown, Quo Warranto, and Assize Rolls.

4 . . Catalogue of Assize Rolls, and Gaol Delivery Rolls.

M 1, 2, 3 Calendars of Patent Rolls and Reports of Royal Commission on Historical Manuscripts, and Cartulaire de la Basse Normandie.

4 . . Transcripts of Royal Letters and various MSS.

M 5 . . King's Silver Books.

N 1—5 ⎫
O 1—5 ⎪
P 1—5 ⎬ King's Silver Books.
Q 1—5 ⎪
R 1—5 ⎭

S 1—4 . Indexes Finium, Hen. VIII. to Victoria.

T 1, 2, 3 . Ditto.

4, 5 . Abstracts of Crown Leases, &c., Agenda Books (L. T. R.)

U 1, 2 . Particulars of Grants, &c., and other Augmentation Office Indexes.

3, 4 . . Indexes to First-Fruits, Tenths, &c.

5 . . Doggett Books (Common Pleas).

V 1—4 . Ditto.

W 1—5 ⎫
X 1—5 ⎬ Palmer's Indexes to Patents and Close Rolls, &c.

Y 1—5 . Close Rolls, Edw. I. to Victoria.

Z 1—5 . Signet Office Indexes, Royalist Composition Papers.

I

B.—CHART OF LONG ROOM, RECORD OFFICE.

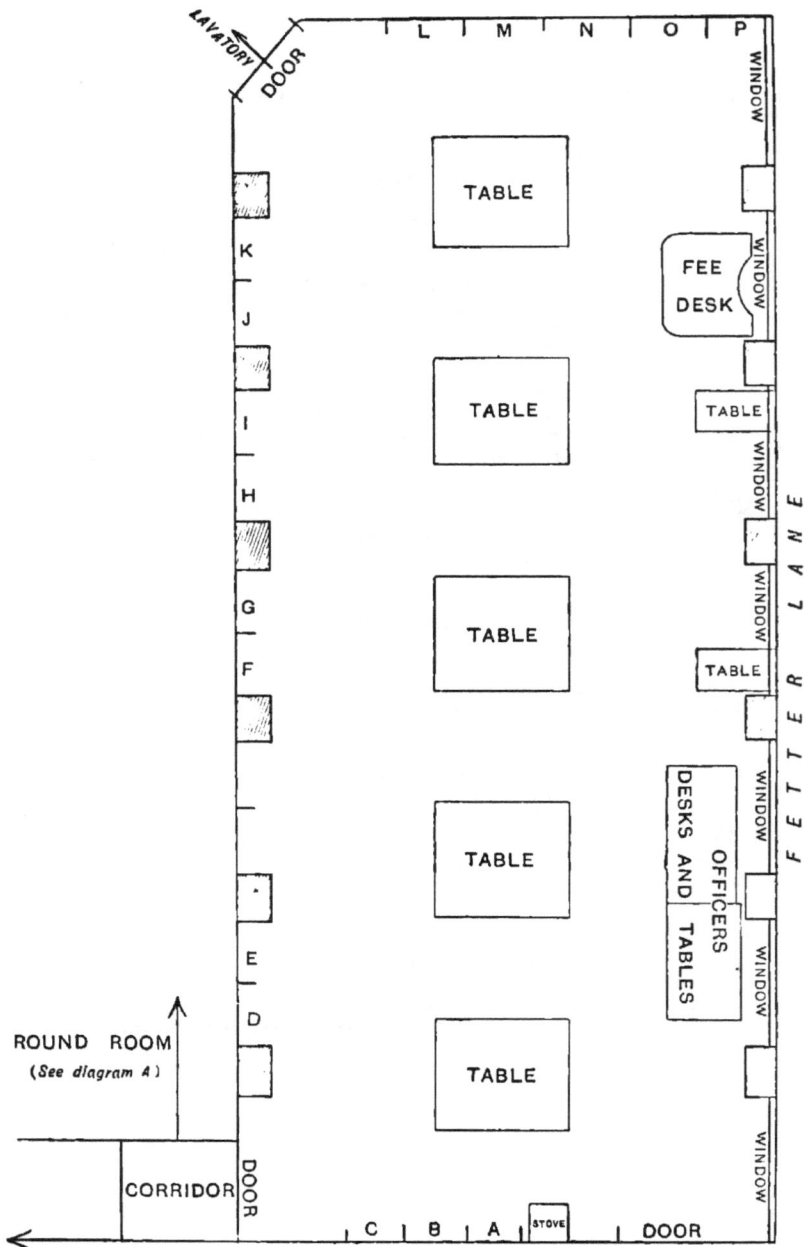

LONG OR LEGAL SEARCH ROOM.—REFERENCES TO CHART OPPOSITE.

S. R. SCARGILL-BIRD, *Superintendent.*

A 1, 2 . . Reports of Deputy Keeper.

3, 4, 5, 6

B 1—6 . . } Bill Books (Chancery).

C 1—6 . . } Petitions (Chancery).

D 1—5 . . } Chancery Pleadings.

E 1—4 . . } Indentures.

F 1—5 . . Recoveries (Common Pleas).

G 1—5 . . Fines (Common Pleas). Deeds Index (Common Pleas).

H 1—5 . . Patent Rolls, Inquisitions Post Mortem, and Parliament Rolls.

I 1—5 . . Duchy of Lancaster Indexes and Acknowledgment of Deeds by Married Women.

J 1—6 . . Exchequer Bills and Decrees, &c.

K 1—6 . Chancery Affidavits.

L 1—6 . . }

M 1—6 . . }

N 1—6 . . } Chancery Decrees and Orders.

O 1—6 . . }

P 1—6 . . Chancery Reports and Certificates.

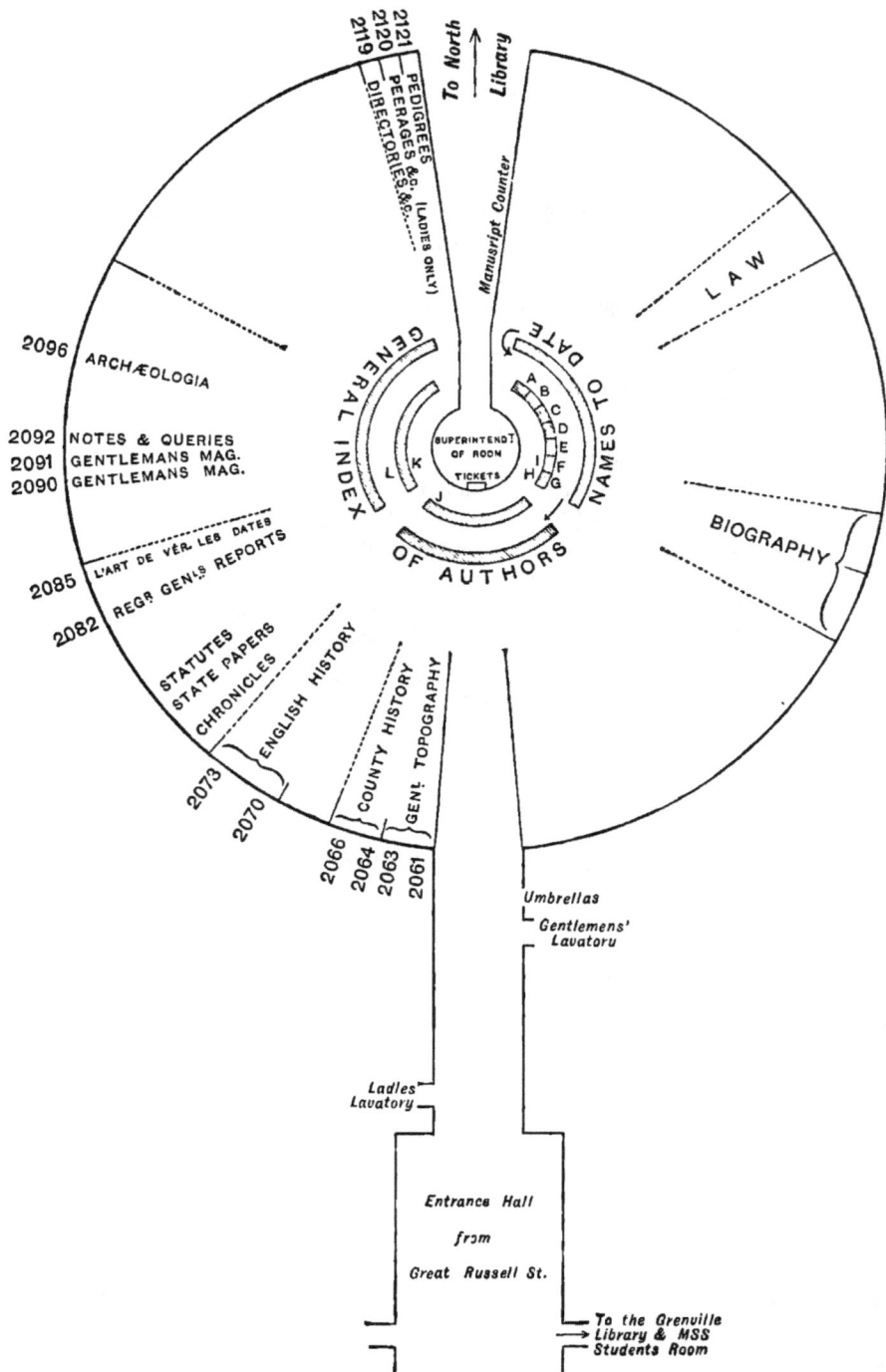

To North
Library
Manuscript Counter

2121 PEDIGREES &c. (LADIES ONLY)
2120 PEERAGES &c.
2119 DIRECTORIES &c.

GENERAL INDEX

NAMES TO DATE

2096 ARCHÆOLOGIA

2092 NOTES & QUERIES
2091 GENTLEMANS MAG.
2090 GENTLEMANS MAG.

2085 L'ART DE VÉR. LES DATES
2082 REGR GEN'S REPORTS

STATUTES
STATE PAPERS
CHRONICLES

2073
2070 ENGLISH HISTORY

2066 COUNTY HISTORY
2064
2063 GEN'L TOPOGRAPHY
2061

SUPERINTEND'T
OF ROOM

TICKETS

A B C D E F G H I J K L

LAW

BIOGRAPHY

OF AUTHORS

Umbrellas
Gentlemens'
Lavatory

Ladies
Lavatory

Entrance Hall
from
Great Russell St.

To the Grenville
Library & MSS
Students Room

BRITISH MUSEUM READING ROOM.—CATALOGUES AT THE DESK,
SHOWN ON THE CHART OPPOSITE.

A . Harleian, Arundel, Cottonian, Lansdowne, and Sloane MSS.

B .

C . } Additional MSS.

D .

E . Index to great classed catalogues of MSS., kept in the MSS. Students' Room.

F . Miscellaneous (chiefly oriental).

G . Printed Pedigrees and other Genealogical Printed Books.

H . Index of Seals; classed catalogues of Additional Charters, with indexes of names and plans.

I . Calendars of Harley, Wolley, Campbell, and Egerton Charters.

J . Deputy Keeper's Reports.

K . Record Publications; Domesday Book, &c.

L . Historical MSS. Commission Reports.

*** It has not been possible in this sketch plan to show the positions of the catalogues A to I satisfactorily. *All* are on the inner side of the desk (*i.e.*, towards the central point of the room), but A to G are on the upper shelf, and H and I on the lower.

THE BRITISH MUSEUM.

The use of the Reading-room is restricted to the purposes of study, reference, and research. The room is kept open on every day of the week except Sunday, and except Good Friday, Christmas Day, and any fast or thanksgiving day appointed by authority; except also the first four week-days of March and October, when the Museum is closed for cleaning.

The hours throughout the year are from nine in the morning till eight in the evening from September to April inclusive, and till seven during the other months.[1]

The following are the regulations in force :—

Persons desiring to be admitted to the reading-room must apply in writing to the Principal Librarian, specifying their profession or business, their place of abode, and, if required, the purpose for which they seek admission.

Every such application must be made two days at least before admission is required, and *must be accompanied by a written recommendation from a householder* (whose address can be identified from the ordinary sources of reference), or a person of recognised position, with full signature and address, *stated to be given on personal knowledge of the applicant, and certifying that he or she will make proper use of the Reading-room.*[2]

If such application or recommendation be unsatisfactory, the Principal Librarian will either refuse admission, or submit the case to the Trustees for their decision.

The tickets of admission must be produced if required, and are not transferable.

No person under twenty-one years of age is admissible, except under a special order from the Trustees.

Readers may not write upon, damage, or make any mark upon any book, manuscript, or map, belonging to the Museum.

Readers may not lay the paper on which they are writing on any book, manuscript, or map.

No tracing is allowed to be made without express permission from the Principal Librarian.

[1] Artificial light not being used in the (except in the Reading-room) Library, books cannot be supplied for the Reading-room service after 3.30 in January, February, November, and December ; 4.30 in March and October; 5.30 in April and September; and 6.30 in May, June, July, and August.

[2] The Trustees cannot accept the recommendations of hotel keepers and lodging-house keepers in favour of their lodgers.

Silence must be [1] strictly observed in the Reading-room.

Readers are particularly requested to replace on the shelves of the Reading-room, as soon as done with, such books of reference as they may have had occasion to remove for the purpose of consultation.

Any reader taking a book, manuscript, map, or other property of the Trustees out of the Reading-room, will be dealt with according to law.

Readers, before leaving the Reading-room, must restore to an attendant at the centre counter, all books, manuscripts, or maps, which they have received, and must reclaim and get back the tickets by which they obtained them. Readers are held responsible for such books, manuscripts, or maps, until the tickets have been re-delivered to them.

Any infringement of these rules will render the privilege of admission liable to forfeiture.

Cases of incivility, of undue delay in supply of books, or other failure in the service, should be immediately reported to the Superintendent of the Reading-room.

The privilege of admission is granted upon the following conditions :—

(*a*) That it may be at any time suspended by the Principal Librarian.

(*b*) That it may be at any time withdrawn by the Trustees in their absolute discretion.

All communications respecting the use of the Reading-room must be addressed to "The Principal Librarian, British Museum, W.C."

It is requested that any reader observing a defect in, or damage to, a book, manuscript, or map, will point out the same to the Superintendent of the Reading-room.

Of the numerous printed Indexes to the contents of the British Museum I can only refer shortly :—

Cottonian Library (Smith), Oxford, folio, 1696
King's Library (Casley), London, 4to., 1734
Sloane, Birch, and other Libraries (Ayscough), 2 vols., London, 1782
Cottonian Library (Planta), London, folio, 1802
Harleian Library, London, folio, 1808-12
Printed Books (Ellis and Baber), 8 vols. 8vo., 1813-19
Hargrave MSS. (Ellis), 4to., London, 1818
Lansdowne MSS. (Ellis), folio, 1819
King's Library (Barnard ?), 5 vols. folio, 1820-9
Maps, Prints, Drawings, &c., in King's Library, folio, 1829
Maps, Plans, and Charts (all Museum), 2 vols. 4to., 1885, by R. K. Douglas. [A splendid work.]

---

[1] But is not.

Arundel MSS., 1834, folio
Burney MSS., 1840, folio
Printed Books only, folio, London, 1841 (letter A only)
MSS. Maps, Charts, and Plans, and Topographical Drawings, 2 vols.
    8vo., 1844
Additional and Egerton MSS., acquired 1783-1835, folio, 1849
Additions to the Museum, acquired 1831, 8vo.
        (*i.a.*) Descriptive Catalogue of Additional MSS. 8220-8901
Additions to the Museum, acquired 1832, 8vo.
        (*i.a.*) Descriptive Catalogue of Additional MSS. 8902-9344
Additions to the Museum, acquired 1833, 8vo.
        (*i.a.*) Descriptive Catalogue of Additional MSS. 9408-9707
Additions to the Museum, acquired 1834, 8vo., 9375-9825
    ,,   ,,   ,,   1835, 8vo., 9827-10,005
    (These last five volumes also include notes of Printed Books,
    Egerton MSS., and Additional Charts.)
Additional MSS., &c., acquired 1836-47, 3 vols, 8vo., 10,464-17,277
    ,,   ,,   1848-53, 1 vol., 17,278-19,719
    ,,   ,,   1854-60, 1 vol., 19,720-24,026⎫   no
    ,,   ,,   1861-75, 1 vol., 24,027-29,909⎭indexes
    ,,   ,,   1854-75, digest and index of last two
    ,,   ,,   1876-81, 1 vol., 29,910-31,896, index
    ,,   ,,   1876-81, 8vo., 1882
From 1836 the vols. have excellent indexes.

The following Hand-books, &c., will be found most useful, as will the coloured map of the Reading-room, sold in the vestibule for 6*d.*, which shows the position of the chief reference books :—

Sims. Handbook to the Library of the British Museum, &c., with
    some Account of the Principal Libraries in London, 12mo.,
    London, 1854
Rye (W. B.) List of the Books of Reference in the Reading-room
    of the British Museum, 8vo., 1859
Nichols. A Handbook for Readers, 8vo., 1866
Anderson. The Book of British Topography : a classified Catalogue
    of the Topographical Works in the Library, relating to Great
    Britain and Ireland, 8vo., London, 1881

The chart on p. 116 will show the position of the reference books and indexes most likely to interest the record searcher proper.

PROBATE REGISTRY.

Department for Literary Inquiry.

This is located in the south side of the great quadrangle of Somerset House, in the basement beneath the General Search-room (where the general public pay their shillings to see any will or administration). The literary enquirer enters by the main entrance to the Principal Registry, and passes through a door on the left, and proceeding through a passage with rooms on the right and left, descends by a staircase on the left hand to the basement, where are also rooms on either side of the corridor.

Entering No. 9 he finds himself in a large apartment, formed by two rooms being thrown into one, and furnished with tables and desks for the use of the literary searchers.

Originally, only six students could be accommodated at one time; but since the autumn of the year 1884, when a second room was added, as many as fourteen persons can pursue their enquiries together.

The use of the calendars, registers, &c., is under the direction of the Superintendent of the Department, Mr. J. C. Challoner Smith, a gentleman ever ready to offer valuable suggestions and assistance to all who are earnest students, and not mere triflers.

There are two messengers to fetch the calendars and registers as they may be wanted. Not more than two registers are permitted to one reader at one time, nor more than eight altogether in one day.

The Literary Department is open from 10 to 3.30 Monday to Friday, and from 10 to 1.30 on Saturdays, except during the long vacation, when the hours are from 10 to 1.30 on Saturdays, and from 11 to 2.30 on other days. It is closed for a period of six weeks during the autumn vacation.

The visitor is allowed, without fee, to search the calendars, to read and make any notes from the registered copies of wills, from the earliest recorded to within a hundred years of the particular year in which he makes his enquiry.

The department was created in 1862, and for many years the

period to which literary enquirers were restricted was from 1394 to 1699; in less than three years after the removal from Doctors' Commons to Somerset House, it was extended to 1 George III. (1760); and in September, 1884, in compliance with a numerously signed petition from literary men, scholars, &c., it was still further extended to a hundred years from the year in which the student makes his search, so that a new calendar and register becomes available upon the 1st January in each year.

Visitors are only permitted to read and inspect the *registered* copies of wills : the originals are not open to them.

During the Cromwellian period there is a gap in the records of all the minor courts; all wills being proved in the London Registry.

The registered copies of wills contained in the Prerogative Court commence with the year 1384.

The Admonition Act Books are complete from the year 1559 to the present time, except that for 1662, which is missing.

The wills contained in the Commissary Court of London begin in 1374, and extend with sundry gaps until 1857.

The books of the Consistory Court of London contain a large number of marriage licenses; also various ecclesiastical proceedings relating to divorce, &c., &c., interspersed among the wills.

In may be mentioned that amongst the records of the Prerogative Court are a large number of inventories (many thousands), which, however, are unfortunately in a chaotic state, and seem likely to remain so. The Principal Registry includes Hertford, South Essex, Middlesex, Surrey, and West Kent.

The districts assigned to the other Registers are those specified in Schedule A of the Probate Acts, 20 and 21 Vict., chap. 17 (1857).

The records of the Minor Courts mentioned above, as included in the Principal Registry are under the charge of Mr. G. H. Rodman, a gentleman of long experience, who is ably assisted by Messrs. Cheyne and Rouse. The writings relating to these minor courts (Episcopal, Archidiaconal, and Peculiar) commence at various irregular dates, some as early as the fourteenth, whilst others do not begin before the seventeenth century.

The enquirer will occasionally find an index extant, whilst the wills for the corresponding period are not in the Registry, never having been transferred to it, but lost by the carelessness or neglect of former registrars in the districts where they were kept in past times.

Besides the above, certain records are preserved here, owing to special circumstances, relating to Berks, Bucks, and Oxford, the Diocese of Salisbury, and the Archdeaconry of Richmond (Yorks.)

The Archdeaconry of Richmond extended over parts of Yorkshire, Lancashire, Westmoreland, and Cumberland.

For the three Eastern Deaneries (Richmond, Catterick, and Borough Bridge), the records come down to 1858.

For the five Western Deaneries (Amounderness, Copeland, Furness, Kendal, and Lonsdale) the records come down to 1748, and after that date are to be found at the Lancaster District Registry.

The following are the official instructions to readers :—

Calendars :—In writing for Calendars [each of which is numbered] it is necessary to give the numbers [and not the dates] of the respective calendars which are required. Readers must replace calendars which they take from shelves in the room.

Registers :—In writing for Will-Register books, the name of the book and the folio must be given [e.g., fo : 73, "Juxon."] No reader is allowed to have more than two of these books at one time, nor more than eight of them in one day.

Act-books :—In writing for Act-books, it must be stated whether Probate-Act books or Administration-Act books are required, and beyond this, it is only necessary to mention the year or years to which they refer thus—" Probate-Act book, 1697," or "Admon-Act book, 1705."

Readers must, on each occasion of their attendance, sign their name in the book provided for that purpose.

The following are the regulations concerning the admission of literary inquirers to the principle Probate Registry at Somerset House :—

1. Application is to be made by letter addressed to the President of the Probate, Divorce, and Admiralty Division of the High Court

of Justice, at "The Principal Probate Registry, Somerset House, London," with "Department for Literary Inquiry" in the corner of the envelope.

2. The applicant is to state his name, address, profession, or description, the object of research, and the period during which he proposes to attend. If considered necessary, he may be called upon for further explanation, or for a reference.

3. An order under the signature of the President will give the applicant free admission for literary purposes during the time specified therein, subject to the requirements of Regulation No. 4. This privilege will be liable to forfeiture for any breach of the rules or regulations, or any injury to, or want of care in the use of, the books or documents.

4. Every visitor will be required on each occasion of his attendance, to sign his name in a book provided for that purpose.

5. The visitor will be allowed without fee to search the calendars of the wills proved and administrations granted at a time not less than 100 years prior to the search being made, to read the registered copies of wills proved, and the probate and administration act books to the same date, and to make extracts from such wills and books.

6. The visitor will not be allowed to trace or take an impression from the writing of any book or document in the Registry, or to use any ink in making extracts.

7. No more than two register books can be produced for one reader at the same time.

8. The Superintendent of the department will arrange the days for the attendance of those who are entitled to admission, and, as far as possible, give facility for each person who has commenced a search and inquiry to complete the same without interruption.

9. The Department for Literary Inquiry in the Probate Registry shall be under the immediate superintendence of the Record Keepers, and shall be open at all times when the Registry is open, except for six weeks in the months of August, September, or October, commencing from a day to be fixed by the Senior Registrar, when it will be closed.

10. No book or document shall be searched for, looked up, or produced on Saturdays after 1.30 p.m., or on other days within the last half hour of the Department remaining open, unless one of the Registrars shall otherwise direct.

Dated the 19th November, 1884.

(Signed) JAMES HANNEN.

CITY OF LONDON RECORDS.

The Records of the *City of London* are under the charge of Mr. Reginald R. Sharpe, D.C.L. (St. John's, Oxford), to whom enquiries as to access, &c., should be made, at the Record-room, Guildhall, E.C.

The numerous printed calendars and works, such as *The Liber Albus*, and the *Memorials of London and London Life, in the Thirteenth, Fourteenth, and Fifteenth Centuries* (Riley), the *Remembrancia*, 1579-1664 (Overall), and the *Calendar of Letters*, 1350-1370 (Sharpe), the publication of which does so much credit to the City of London, are referred to elsewhere. A much finer edition of the *Liber Albus* and the *Liber Custamarum* has been printed in the Rolls Series, see p. 156.

LAMBETH LIBRARY.

I am indebted to Mr. S. W. Kershaw, the able and talented librarian of this Library, for permission to reprint the following excellent account of it by him :—

The MSS., in all some 1300 volumes, are divided into seven series, named after their respective donors :

1. LAMBETH MSS. (Nos. 1-576).—Given by several Archbishops.
2. WHARTON MSS. (577-596).—Those of Henry Wharton, purchased by Archbishop Tenison.
3. CAREW MSS. (596-638).—Those formerly belonging to Lord Carew, purchased by Tenison.
4. TENISON MSS. (639-923).—Collected and given by Archbishop Tenison.
5. GIBSON MSS. (929-942).—Formerly belonged to Archbishop Tenison, who gave them to his librarian, Edmund Gibson, afterwards Bishop of London, by whom they were deposited at Lambeth.
6. MISCELLANEOUS MSS. (943-1174).—Presented by various benefactors.
7. MANNERS-SUTTON MSS. (1175-1221).—Those purchased and presented by Archbishop Manners-Sutton.

In 1868 the Library was endowed by the Ecclesiastical Commissioners, and certain alterations effected by the late Archbishop Longley, and approved by the present Primate, have rendered the contents easily accessible for research.

It may be interesting to add that the collection consists of nearly 30,000 volumes, which were formerly arranged in the galleries over the once standing cloisters. The books are now placed in the Great Hall, rebuilt by Archbishop Juxon about 1661, and very suitably arranged for the purpose about 1828, at the cost of Archbishop Howley. The roof of the hall, which is of noble dimensions, and resembles those of Westminster Hall and Hampton Court Palace, is built of English oak, carved on several parts of the woodwork with the arms of Juxon, and of the See of Canterbury. At one end of the hall is a window in which are the armorial bearings of many of the archbishops, together with portions of stained glass, which have been removed from other parts of the palace to this window.

The Lambeth Collection consists of records, MSS., and printed books. The two former are made known in print by the catalogue compiled by Dr. H. J. Todd, and published in 1812 (folio).

The records include,—

1. *Registers of the Archbishops of Canterbury,* from Archbishop Peckham, 1279, to Potter, 1747, and are rendered highly valuable by Ducarel's elaborate indexes.

There are indexes of *all* these records (in MS.), either in separate books or *pre*fixed to each volume.

It may be hardly needful to state that these volumes contain, in general, the account of each archbishop's consecration, records of ordinations, visitations, institutions to benefices, a number of important wills, proceedings with the suffragans, and Convocation, and, in fact, all the most important proceedings of the primate.

2. *The Parliamentary Surveys,* circ. 1650, in 21 volumes, contain surveys of livings taken at that time.

3. *Augmentations of Livings.* (Nos. 966-1021.) These

papers relate to "salaries or pensions made by ordinance of Parliament for maintenance of preaching ministers from 7th Feb., 1647, to 25th Dec., 1658."

4. *Cartæ Antiquæ.* 13 vols. (Nos. 889-901.) Certain "charters and instruments relative to the See of Canterbury, and others within that province. Some of these instruments are of ancient date, but most of them are of Henry VIII.'s reign, and subsequent thereto."

5. *Presentations to Benefices.* 4 vols. (Nos. 944-7.) Made during the Commonwealth.

6. *Leases.* 3 vols. (Nos. 941-50.) These are counterparts of leases of Church lands, made by trustees, under authority of Parliament, 1652-58.

7. *Notitia Parochialis.* 6 vols. (Nos. 960-65.) These returns give an account of the state of 1579 parish churches in the year 1705.

8. *Surveys of the Possessions of the See of Canterbury and of Peculiars.* (3 vols.) These are kept separate from the possessions of the other sees, deans, and chapters. The volumes contain some original surveys, and some transcripts.

The above summary gives the titles of the Lambeth "Records," as distinguished from the MSS.[1]

The following are the regulations now in force :—

REGULATIONS.

1. With the exception of the periods named in regulation No. 2, the library is open to the public on Mondays, Wednesdays, Thursdays, and Fridays from 10 a.m. to 4 p.m.; from April to July (both months inclusive) until 5 p.m.; and during the forenoon of Tuesdays.

2. The library is closed during the week commencing with Easter Day, for seven days computed from Christmas Day, and for a period of six weeks, from about the first day of September in every year.

---

[1] Mr. Kershaw then describes the various ecclesiastical, manorial, heraldic, historical, and antiquarian MSS. contained in the library, not being "records" themselves.

3. Extracts from the MSS. or printed books are allowed to be made freely, but in case of a transcript being desired of a *whole* MS. or printed book, the consent of the archbishop must be previously obtained.

4. Permission to copy illuminated MSS. and rare engravings, can only be obtained on submission of the applicant's name to the Archbishop.

5. MSS. are only lent out by an order signed by the Archbishop, and with a bond of £50 or £100 for their return within six months or on demand.

The foregoing Regulations have been approved by the Ecclesiastical Commissioners for England, in pursuance to an Order in Council, June, 1880.

Rules for Lending Books.

1. Except under special permission, the loan of books will be restricted to the clergy and laity of the diocese of Canterbury, and to persons residing within the parish of Lambeth, the borough of Southwark, and the city and liberties of Westminster.

2. All applications for the loan of books must be countersigned by a *beneficed* clergyman, who must certify to his *personal* knowledge of the applicant.

3. Every one desiring to borrow must obtain from the Librarian the prescribed forms, and all applications for the removal of books must be made to the Librarian in person, or by some trustworthy representative, *authorized in writing by the applicant, to receive them.*

4. Except by special permission, not more than two volumes to be borrowed at the same time, and such volumes must be returned to the Librarian within two months from the date of removal.

5. If any volume or volumes of a set be defaced, injured, or lost, such volume or volumes shall be replaced by others of equal value and in similar condition.

6. Works of reference, books of prints, works of an earlier date than 1700 A.D., pamphlets and such books as in the discretion of the Librarians cannot easily be replaced, can only be consulted in the library.

7. Every one borrowing books shall sign, on behalf of himself or the person he represents, a declaration assenting to the foregoing rules; books can be lent out and returned only during library hours.

THE HERALDS' COLLEGE.

This building is situate in Queen Victoria Street, and was formerly entered from Bennet's Hill, Doctors' Commons. It is the sole remaining public office supported entirely by fees; the salaries of its thirteen officials being nominal, and amounting in the aggregate to little more than £200 per annum. Here is preserved the largest and most valuable genealogical and heraldic collection in this or in any other country in the world, the College having been incorporated by charter from King Richard III. There are two classes of documents preserved here : the Records and the Collections. The former comprise—(1) the series of books called Visitation Books, containing the pedigrees and arms of the nobility and gentry of the kingdom, from 21 Henry VIII. to the latter end of the seventeenth century, taken under royal commission, the first being issued in 21 Henry VIII. and the last in 2 James II. (2) Books of Modern Records. These contain the miscellaneous pedigrees of the nobility and gentry which have been recorded since the discontinuance of the Visitations. (3) Books of Pedigrees and Arms of Peers, compiled pursuant to the standing orders of the House of Lords of 11th May, 1767. (4) Books of Pedigrees and Arms of Baronets, under a royal warrant of 3rd December, 1783, "for correcting and preventing abuses in the Order of Baronets." (5) Funeral Certificates. These contain attested accounts of the time of death, place of burial, and of the marriages and issue of the several members of the nobility and gentry whose funerals were attended by the Officers of Arms or their deputies. (6) Books containing accounts of Royal Marriages, Coronations, Funerals, &c., &c. (7) Books called "Earl Marshal's Books," from the time of Queen Elizabeth, containing entries of such instruments and warrants under the royal sign manual as relate to the arms of the blood royal, licenses from the crown for change of name and arms, or for acceptance of foreign honours, &c., &c. (8) Grants of Arms. These comprise the grants of armorial bearings down to the present day.

K

The above are the official records of the College, made by the proper officers in the regular exercise of their duty, and are admissible as evidence in the courts of law.

Of the second class of documents, "the Collections," it is only necessary to state that they consist of some two or three thousand volumes of manuscript, the accumulated labours of Glover, Camden, Vincent, Philpot, Dugdale, Le Neve, Walker, St. George, Harrison, Heard, Warburton, Townshend, Brooke, Anstis, Bigland, Courthope, Beltz, Pulman, Young, King, Collen, and other distinguished and skilful members of the College, to say nothing of purchases, such as the collection of Parish Registers of the late Col. Chester, &c., &c., &c.

Access to the College is much easier and the fees much lighter than people generally suppose. An ordinary search upon a personal application is 5s. ; if by correspondence, the fee is 10s. 6d ; a general search through the records, £2. 2s. ; and a general search through the records and the collections, £5. 5s.— fees by no means out of proportion, when the time and the trouble involved is taken into consideration. Transcripts of pedigrees are charged 5s. for each generation shown on the pedigree, an extra charge being made for a sketch of the arms, which charge, of course, varies according to the number of quarterings, &c.

An application to the College is made in one or the other of two modes. A perfect stranger would present himself in the public office, where he would find the herald and pursuivant, who happen to have, in rotation, the turn of waiting for that month, and the fact of his application to them gives them the sole right to the transaction of the business. A person, on the other hand, who has a knowledge of an individual officer of the College, either on the score of personal acquaintance or through the recommendation of a common friend, would make his application to that officer in his private apartment, who, in like manner, would have a right to the transaction of his particular business in the same manner that the professors of law, medicine, &c., &c., have. But a searcher will probably know the name of some herald or pursuivant as being himself interested in the county or district whence he comes, *e.g.*, Mr.

Athill is specially conversant with the Eastern Counties, and in that case it is far best to go direct to the specialist.

One thing I can say, of my own knowledge, that the officers are far more conscientious and careful than they used to be, and now-a-days very seldom pass fudged pedigrees, as formerly was the case. I believe the number of absurd questions put to the officers in charge passes all limit. Of course, one can readily understand that the new man, whose successful trading has taught him he can—or ought to—get anything by paying for it, should expect that, on payment here, a perfect pedigree, Norman for choice, can be turned out for him on demand, without the necessity of any research ; but I must own to being amused the other day on hearing that the august authorities are not infrequently addressed by children and other collectors of "coats of arms" and "crests" from envelopes, who seem to consider the College a sort of wholesale and retail emporium of stamped stationery.

The inclusive charge for a Grant of Arms is £76. 10s., including all fees. Some people who want to change their name under the directions of a will or settlement, are under the mistaken idea that they must get the Royal License or an Act of Parliament to enable them to do so. This is not the case, as it is well-established law that a man may change his surname of his " mere motion " (though it is wiser to do so by deed enrolled in Chancery for safe custody, supplemented by an advertisement in *The Times*), but if anyone has the desire to waste money in a lordly way, he can do so through the College for about £100 : both the War Office and the Civil Service, however, now recognising the enrolment and advertisement. The mistake probably arose from the fact that a royal license *is* necessary to adopt the arms of another family, and the arms and name are generally changed by the same instrument.

I have elsewhere referred to the Calendar of the Arundel MSS. belonging to the College, which was printed by Sir C. G. Young. This is. I think, the only printed catalogue of MSS. lodged here.

No catalogue or index has ever been published of the contents of the College, nor have the Officers of Arms ever considered themselves called upon to issue, as they most certainly ought to do, a list of those persons who are, and those persons who are *not*, entitled to armorial bearings. Many of the earlier Visitations have been copied, more or less accurately, rather less than more, and are to be found in the British Museum and other libraries. The greater number of them have been printed by the Harleian and other societies. Of these, I think the following is a fairly correct list:—

Printed Visitations.[1]

[For details as to these prints, and as to the unprinted Visitations, see two very valuable articles by Dr. Marshall (Rouge Croix) in *The Genealogist*, new series, i. p. 201 ; ii. pp. 148 and 461, and iii. pp. 35 and 112, from which this Appendix is chiefly compiled.]

BEDFORDSHIRE.—1566, 1586, and 1634, printed for the Harleian Society, vol. xix., by F. A. Blades.

BERKS.—1566, printed in *The Genealogist*, by W. C. Metcalfe, who also printed the Visitation of 1664-6 in the same paper. Both have been reprinted separately.

BUCKS.—1566, printed in *The Genealogist* by W. C. Metcalfe; also separately printed.

CAMBRIDGE.—1619, printed by Sir T. Phillipps in 1840. 1684, portion of, printed in *The Genealogist*.

CHESHIRE.—1533, 1566-7, and 1580, printed for the Harleian Society, vol. xviii., by J. P. Rylands.

CORNWALL.—1530, 1573, and 1620, printed by Col. Vivian. 1620, printed for the Harleian Society, vol. ix., by Col. Vivian and H. H. Drake ; also printed, from A to C, by Sir H. Nicolas in 1838.

---

[1] **Berry's** series of *County Genealogies*, issued about half a century ago, should not be overlooked. They are as follows:—Sussex, 1830; Kent, 1830; Hants, 1833 ; Surrey, 1837 ; Berks, 1837 ; Bucks, 1837; Essex, n.d. ; Herts, n.d. Sir T. Phillipps' folio vol. of printed pedigrees, 1850, is also noticeable, as is his privately printed *Index to the Visitations* in his library, 1841.

CUMBERLAND.—1530, printed for the Surtees Society, vol. xli., by W. H. D. Longstaffe. 1615, printed for the Harleian Society, vol. vii., by J. Fetherston.

DERBYSHIRE.—1662-4, printed in *The Genealogist;* also separately issued.

DEVONSHIRE.—1531, 1564, and 1620, now being printed by Colonel Vivian. 1620, printed by F. T. Colby, 8vo., 1872, Harleian Society, vol. vi. 1564, printed by F. T. Colby, 8vo., 1881. 1620, also partially printed, with additions, by J. Tuckett in 1863.

DORSET.—1565, printed in *The Genealogist* by W. C. Metcalfe; also separately issued 1887. 1623, printed for the Harleian Society, vol. xx., by J. P. Rylands.

DURHAM.—1530, printed for the Surtees Society (xli.) by W. H. D. Longstaffe. 1575, printed by N. J. Phillipson, 1820, fo. 1615, printed by Sir C. Sharp and J. B. Taylor, 1820, fo. 1575, 1615, and 1666, printed by J. Foster, 1887, 8vo.

ESSEX.—1552, 1558, 1570, 1612, and 1634, all printed by W. C. Metcalfe, 1878, 8vo., Harleian Society, vols. xiii. and xiv. 1664-8, now being printed by J. J. Howard.

GLOUCESTER.—1623, printed for the Harleian Society, vol. xxi., by Sir J. Maclean and W. C. Heane. [As far as possible the Visitations of 1569 and 1583 have been worked into this.] 1682-3, printed by T. F. Fenwick and W. C. Metcalfe, 1884, 8vo.

HEREFORDSHIRE.—1569, printed by the Rev. F. W. Weaver, 1886.

HERTS.—1572, 1634, printed for Harleian Society, vol. xxii., by W. C. Metcalfe.

HUNTS.—1613, printed for the Camden Society by Sir H. Ellis.

KENT.—Some printed by Berry. 1619, begun by the Kent Archaeological Society (J. J. Howard).

LANCASHIRE.—1533, printed by the Chetham Society by W. Langton. 1567, 1613, and 1664-5, printed by the Chetham Society by F. R. Raines.

LEICESTER.—1619, printed for the Harleian Society, vol. ii., by J. Fetherston.

LINCOLNSHIRE.—1562-4 and 1592, printed in *The Genealogist* by W. C. Metcalfe ; also separately issued in 1882.

LONDON.—1568, printed for the Harleian Society, vol. i., by J. J. Howard and G. J. Armytage ; also London and Middlesex Society, a few pedigrees annotated. 1633-5, printed by the Harleian Society, vols. xv. and xvii., by J. J. Howard and J. L. Chester.

MIDDLESEX.—1663-4, printed by Sir T. Phillipps, 1820, fo. Mr. Foster has issued another edition of this, with Sir G. C. Young's notes, 1887, 8vo.

NORFOLK.—1563, begun to be printed by the Norfolk and Norwich Archæological Society very many years ago, but is now being issued at the rate of a few pages per annum. 1563, 1589, and 1613, preparing to be printed for the Harleian Society by W. Rye. An index has been printed by Mr. Athill to the last *Norfolk Visitation*.

NORTHAMPTON.—1564, 1618-9, printed by W. C. Metcalfe.

NORTHUMBERLAND.—1615, printed in *The Genealogist* by Dr. Marshall ; also separately issued in 1878.

NOTTINGHAM.—1569, 1614, printed for the Harleian Society (vol. iv.) by Dr. Marshall.

OXFORD.—1566, 1574, and 1634, printed for the Harleian Society, vol. v., by W. H. Turner.

RUTLAND.—1618, printed for the Harleian Society, vol. iii., by G. J. Armytage.

SOMERSET.—1531, 1573, and part of 1591, printed by F. W. Weaver, 1885, 8vo. 1623, printed for the Harleian Society, vol. xi., by F. T. Colby; also partially printed by Sir T. Phillipps, 1838.

STAFFORD.—1583, 1614, and 1663-4, printed for the W. Salt Society, by H. S. Grazebrook. 1663 (an abstract), printed by Sir T. Phillipps, 1854.

SUFFOLK.—1551, 1577, 1611, printed by W. C. Metcalfe. 1561, 2 vols., containing only a few pedigrees annotated, printed by Dr. Howard; never completed or indexed.

SURREY.—1623, begun to be printed by the Surrey Archæological Society in 1858, but that society seems about as prompt in its issue as the Norfolk society. An index to the pedigrees printed has been privately issued by W. C. Metcalfe.

WARWICK.—1619, printed for the Harleian Society, vol. xii., by J. Fetherston; and partially in the *Warwickshire Magazine*, 8vo. 1859; (selections from), printed by the Rev. G. H. Dashwood at his private press, 1865, very scarce; and also partially in the *Warwickshire Antiquarian Magazine*, cf. *Genealogist* (n. s.), iii. 36.

WESTMORELAND.—1530, printed for the Surtees Society, vol. xli. 1615, printed by J. G. Bell, Bedford Street, Covent Garden, 8vo., 1853.

WILTS.—1623-77, printed by Dr. Marshall, 1882, 8vo. A portion of the Visitation of 1677 was printed by Sir T. Phillipps, 1828, folio.

WORCESTERSHIRE.—1682-3, printed by W. C. Metcalfe, 1883. Additions to this in the *Midland Antiquary*, and separately, 22 pp. 8vo.

YORK.—1530, printed for the Surtees Society, vol. xli., by W. H. D. Longstaffe. 1563-4, printed for the Harleian Society, vol. xvi., by C. B. Norcliffe, 1881. 1584-5 and 1612, by J. Foster. 1584-5 and 1612, by J. Foster, 1875. 1666, printed for the Surtees Society, 1859, vol. xxxvi. 1860, 8vo., and Index by G. J. Armytage, 1872, 8vo.

Most of the above, however, have been printed from transcripts, and in several cases from incorrect and incomplete transcripts, so reference should always be made to the original Visitations in the College of Arms. Besides those printed there are many other MS. Visitations. A list of those not yet printed, by Dr. Marshall, will be found in *The Genealogist* (n. s.), iii. 115.

Mr. Rylands has published a work, called *Disclaimers at the Heralds' Visitations*, which is of great interest; but the reader must not conclude that all the four thousand odd who disclaimed had *not* a legal right to arms, for in many cases the disclaimer was made to save the Heralds' fees, or from political reasons, or because the person had evidently been surnamed in error.

The Funeral Certificates, already referred to (p. 82), are very valuable records, and give a mass of information about the family of the buried man; an extensive series is amongst the records in the Heralds' College.

THE BODLEIAN LIBRARY, OXFORD.[1]

The Bodleian is open at 9 a.m. throughout the year, closing at 3 p.m. in January, 4 p.m. in February and March, 5 p.m. from April to July inclusive, 4 p.m. in August and to the end of October, and again at 3 p.m. in November and December. When there is a sermon before the University, some four or five times a year, it is not open till 11. It is closed on Sundays, on January 1st to 6th, Good Friday to end of Easter week, Ascension Day, Whit Monday and Tuesday, Commemoration Day, on October 1st to 7th, November 7th and 8th (or 6th and 7th when the 8th is a Sunday), and from Christmas Eve to the end of the year.

Attached to the Bodleian is the Radcliffe Camera, which is, in effect, a separate reading-room in a building close to the Bodleian. This is open from 10 a.m. to 10 p.m. on all days except Sundays, the four days next before Easter, the three days ending on the first Saturday in July, the three days ending on the last Saturday in September, and on Christmas Day and three adjoining week days.

Anyone working at the Bodleian, say in January, can give up his printed books or MSS. or charters to an attendant at 3 p.m. They will be carried across to the Radcliffe, and he can then work up to 10 o'clock at night.

A letter to Bodley's librarian, asking that a particular MS. might be sent over to the Radcliffe, has enabled at least one student to run down to Oxford by an afternoon train, finish a severe piece of collation, and return to London the same night.

All graduates of Oxford have the right of entry, but other

---

[1] For reports on other Oxford MSS., see p. 162.

persons are admitted to study on presenting a satisfactory recommendation. As to this and all other matters the authorities are exceedingly liberal and courteous.

Turner's printed catalogue of Bodleian Charters has already been noticed. The MS. catalogue of the Library[1] fills nearly 800 vols. Admirable catalogues of the Tanner MSS., Ashmolean MSS., and Rawlinson MSS. have recently been printed, and can be had.

For a calendar of the Clarendon State Papers see p. 60; and see Dodsworth in Index.

CAMBRIDGE[2] UNIVERSITY LIBRARY.

This is subject to a series of cramped and illiberal rules. Persons, not being members of the university, desiring tickets of admission to the library, are required to present to the Syndicate letters from two members of the Senate, certifying that the applicant is known to them to be a student in some *specified* subject, and is a fit and proper person to be admitted to the library.

A person thus favoured is permitted to use the library between the hours of 9 a.m. and 1 p.m. on Saturdays, and 10 a.m. and 2 p.m. on other days, except on the days when the library is re-opened or is closed for any quarter. Permissions do not entitle the holder to have access to the locked cases of very rare and early printed books, and the permits expire on the 20th of October of each year.

The library is closed on the following days:—December 24th to 28th inclusive, Ash Wednesday, from the Thursday

---

[1] While at Oxford the student should remember the Ashmolean Library (see Sims, 446). and look at Coxe's printed catalogue of MSS. at the different colleges and halls, 2 vols. 4to., Oxford, 1852. There are also separate calendars of the MSS. of several colleges, *e.g.*, All Soul's, which has also published an admirable volume of its archives.

[2] For reports on other Cambridge MSS., see p. 161.

before to the Tuesday after Easter inclusive, Ascension Day, Monday after Whitsun Day, Sundays and Public Fasts or Thanksgivings, 15th to 30th of September inclusive, two days after the other three Quarter Days.

A very poor catalogue of the MSS. in the University Library was published at the University Press in 1846, in 5 vols. 8vo., with an index by Luard in 1857. The Parker MSS. in the catalogue of Corpus Library, by Stanley, 1722, fo., and St. John's, by commission, 1843, were calendared by Nasmith, 4to., 1727, the Caius MSS., by Smith, 8vo., 1849. An excellent index to the Baker MSS. was published by Macmillan in 1848, 8vo.

*** A few words as to the antiquarian booksellers may not be out of place. *Facile princeps* for important MSS. and rare books is Bernard Quaritch of 15, Piccadilly, who though necessarily high in his prices is more a librarian than a bookseller, for he will often freely allow access to his treasures, and sometimes (*crede experto*) even lend them to a non-buyer. J. Sage of Newman's Row, Lincoln's Inn, makes a speciality of Record books, and has always a large stock of them and other antiquarian works, of which he issues periodical catalogues. Ridler of 45, Booksellers' Row, follows respectfully in Quaritch's wake, and his catalogues, which are low priced, sell out speedily, so should be attended to at once. Willis and Sotheran, and Reeve's and Turner, both of the Strand, also deal largely (*inter alia*) in antiquarian works. For County Histories in good bindings, Bain of 1, Haymarket; Toovey of 177, Piccadilly; and Walford of 320, Strand, are renowned, while bargains may often be picked up from the catalogues of D. Nutt, 270, Strand; Bull and Auvrache, 35, Hart Street; U. Maggs, 159, Church Street, Paddington; Russell Smith of Soho Square; H. Gray, 47, Leicester Square; Burn and Oates (Roman Catholic), Orchard Street, W.; and J. Nield, 14, Great Russell Street.

A list of country booksellers, who deal in old county books, *e.g.*, Golding of Colchester, Hunt of Norwich, &c., would be very useful.

APPENDIX I.—FORMS.

Form of Writ of Diem Clausit Extremum.[1]

.... Dei gratia Angliæ Scotiæ et Hiberniæ Rex [Fidei Defensor], &c., escaetori suo in comitatu .... salutem.

Quia A B de .... qui de nobis tenuit in capite diem clausit extremum, ut accepimus, Tibi præcipimus quod omnes terras et tenementa de quibus idem A B fuit seisitus in dominico suo, ut de feodo, in balliva tua die quo obiit, sine dilatione capias in manum nostrum, et ea salvo custodiri facias donec aliud inde præceperimus et per sacramentum proborum et legalium hominum de eadem balliva tua, per quos rei veritas melius sciri poterit, diligenter inquiras quantas terras et tenementa prædictus Willielmus tenuit de nobis in capite, tam in dominico quam in servitio, in dicta balliva tua dicto die quo obiit, et quantum de aliis et per quod servitium, et quantum terræ et tenementa illa valeant per annum in omnibus exitibus, et quo die idem A B obiit, et quis propinquior hæres ejus sit et cujus ætatis. Et inquisitionem inde distincte et aperte factam in cancellariam nostram sub sigillo tuo et sigillis eorum per quos facta fuerit sine dilatione mittas et hoc breve.

Teste meipso apud Westmonasterium .... die .... anno regni nostri, &c.

Form of Inquisition Post Mortem.

[.........[2]] Inquisitio indentata capta apud .... in comitatu prædicto .... die .... anno Regis domini nostri .... Dei gratia Angliæ Franciæ et Hiberniæ Regis [Fidei defensoris.] &c. ....
Coram .... armigero escaetore dicti domini Regis in comitatu prædicto virtute brevis ejusdem domini Regis de diem clausit extremum, post mortem A B eidem escaetori directi et huic inquisitioni

---

[1] For a very fine and most valuable collection of almost every form of deed see Madox's *Formulare Anglicanum.*

[2] Name of county.

annexi per sacramentum E F, G H, &c., proborum et legalium hominum comitatus prædicti. Qui dicunt super sacramentum suum quod prædictus A B in dicto brevi nominatus in vita sua fuit seisitus in dominico suo ut de feodo de et in manerio de X cum pertinentiis jacente in .... in comitatu prædicto, ac de et in advocatione ecclesiæ de Y cum suis pertinentiis in comitatu prædicto.

Et sic seisitus existens, idem A B in vita sua scilicet ...... die ...... anno domini .... apud .... prædictam, condidit testamentum et ultimam voluntatem suam in scriptis et per eadem dedit et devisavit ....

Et juratores prædicti ulterius dicunt super sacramentum suum prædictum quod prædictus A B de manerio et advocatione prædictis cum suis pertinentiis in forma prædicta seisitus existens, obiit apud .... die .... ultimo præterito ante diem captionis hujus inquisitionis et quod C B est, et tempore mortis prædicti A B patris sui fuit, filius et proximus hæres prædicti A B et ætatis ...... annorum .... mensium et .. .. dierum et non amplius.

Et insuper juratores prædicti dicunt super sacramentum suum prædictum quod prædictum manerium de X cum pertinentiis tenetur, et tempore mortis prædicti A B tenebatur, de domino Rege nunc in capite per servitium .... pro omnibus servitiis. Et valet clare per annum in omnibus exitibus, ultra reprisas .... libras.

Et ulterius juratores prædicti dicunt super sacramentum suum prædictum quod prædictus A B nulla alia sive plura maneria, terra, tenementa aut heredimenta habuit seu tenuit in possessione reversione sive usu tempore mortis suæ in comitatu prædicto, ad notitiam juratorum prædictorum.

In cujus rei testimonium uni parti hujus inquisitionis penes præfatum escaetorem remanenti, tam prædictus escaetor quam juratores prædicti sigilla sua apposuerunt, alteri vero parti ejusdem inquisitionis, penes primum juratorum prædictorum remanenti, prædictus escaetor sigillum suum apposuit.

Data die et anno primo supradictis.

Form of Testament.

In Dei nomine, Amen. Ego A B de C D, sanæ mentis et memoriæ [or, "sana mente ac bona memoria existens;" or, "compos mentis et in sanitate constitutus;" or, in bona et sana memoria mea;" or, "compos mentis et sanæ memoriæ"] condo testamentum meum in hunc modum.

Imprimis, commendo (or lego) animam meam Deo Omnipotenti et

Beatæ Mariæ Virgini et omnibus sanctis. Corpusque (*or*, et corpus) meum ad sepeliendum in ecclesia de .... (*or*, Corpusque meum ecclesiasticæ sepulturæ).

Item lego summo altari prædictæ ecclesiæ pro decimis et oblationibus (meis) oblitis seu minime persolutis.

Item ad fabricationem dictæ ecclesiæ unum novum vestimentum.

Item lego ad novum vestimentum integre emendum summo altari ecclesiæ de ....

Item ad adquisitionem novæ imaginis de Sancto ....

Item ad fabricationem ecclesiæ de ...

Item ad reparacionem ecclesiæ de ....

Item (ad opus) fabricæ (ecclesiæ) de ....

Item ad faciendum unam fenestram ....

Item [bells].

Item volo quod unus capellanus idoneus sustentetur . . ad celebrandum pro anima mea (et pro animabus parentum et antecessorum meorum).

Item lego ad celebrandum pro anima mea, etc.

Item ad gildam de Sancto ....

Residuum vero omnium bonorum meorum (mobilium et immobilium), catallorum et debitorum superius non legatorum .... do et lego (*or*, committo) administrationi executorum meorum ad vendendum recipiendum et disponendum [ad solvendum debita mea] ut ipsi ordinandum et disponendum pro me et anima mea (*or*, pro salute animæ mei et omnium fidelium defunctorum in operibus caritativis) prout ipsi viderent (*or*, velint) Deo melius placere et animæ mei prodesse (*or*, et dictis animis celerius expedire).

Probate Act.

[suprascripti defuncti]
PROBATUM fuit testamentum   or   coram ...
suprascriptum
apud ... die mensis .... juramento .... executoris in hujusmodi testamento nominati ac approbatum et insinuatum.

Et commissa fuit administratio omnium et singulorum bonorum et debitorum (jurium et creditorum) dicti defuncti præfato executori de bene et fideliter administrando ac de pleno et fideli inventorio citra ...... proximo futurum exhibendo, necnon de pleno et vero compoto reddendo ad Sancta Dei evangelia jurato. [Reservata potestate alteri executori in hujusmodi testamento nominato cum venerit]......

....

Specimen of a Nuncupative Will.

In the name of God, Amen, the twentie daye of November Anno Domini 1589, James Tompsone late of the parishe of St. George at Colgate in Norwich, beinge sicke in bodye but of good and perfect memorye (thancks be given to God) made his last will and testament nuncupative in manner and forme followinge and in the presens and hearinge of the wytnes underwritten.

First he beinge asked by one Thomas Parker the elder howe he wolde dispose of his goods and what sisters he had, he answered sayeing that he had three sisters, of whome two he knewe where they dwelled unto eyther of whome he gave xxs. a peece but the thirde sayde he I knowe not where she is, or whether that she be livinge or deade. And then beinge asked agayne who should have the rest of his goods (by the saide Thomas Parker) he sayd Frances ther, meaninge Frances Parker beinge his keper in all his sicknes and then ther present whome he intended then to have marryed they beinge bothe assured together, and havinge then bought and mad ther mariadge apparell; but that God prevented it by his sayde sicknes and deathe. And this is the whole substance and effect of his saide disposition of all his goods which he often repeated before these beinge wittnes thereto. Thomas Parker the elder, Thomas Parker the yonger, and Hellen Pollerd, with other more. This will was proved 28 November, 1589, and administration was granted to Frances Parker.

Form of Fine.

1.—Hæc est finalis Concordia facta in Curia Domini Regis apud . . . in . . . anno regni Regis . . .

2.—Coram . . . . et aliis fidelibus domini Regis tunc ibi præsentibus . . .

3.—Inter A B, petentem, et C D, tenentem [per E F, positum loco ipsius C D, ad lucrandum vel perdendum] . . .

4.—De . . . .   In terra. in prata, in pascuis, in molendinis, in stagnis, et in omnibus pertinentiis ejusdem terræ.

5.—Unde placitum fuit inter eos in Curia domini Regis, scilicet | Unde recognitio de morte antecessoris summonita fuit inter eos in præfata Curia, viz.,

---

6.—Quod idem[1] C D, concessit eidem A B, et heredibus suis totam terram, etc. | Quod idem[1] C D, quietum clamavit pro se et heredibus suis totum jus et clamium qd. habuit in tota terra, etc. | Quod idem[1] C D, recognovit totam terram, etc., esse jus et hereditatem ipsius A B [ut illa quæ idem A B, habuit de dono prædicti C D] et terram illam quietum clamavit a se et heredibus suis imperpetuum.

[The appurtenances are here sometimes set out at length.]

7.—Prædicto A B, et heredibus suis imperpetuum. | Et assignatis suis præterquam viris religiosis.

8.—Et pro hac

9.—Concessione | recognitione | remissione | quieta clamantia | fine et concordia.

10.—Prædictus A B.

11.—Dedit prædicto C D.

.. solidos esterling .. marcas argenti .. besantia ..denarios. | unam juvencam ... unum ostorium sorum

Concessit predicto A B, totam vestituram terræ pro annis.

Concessit predicto et heredibus suis C D.

Quietum clamavit imperpetuum totam terram quam tenuit de illo in X. predicto et heredibus suis C D.

Concessit predicto CD, et heredibus suis ... acras predictæ terræ scil.. [or totam predictum terram] (in subinfeudation as then mentioned).

12.—Et prædictus C D, et heredes ejus waramizabunt eidem A B, prædicta tenementa, etc., cum pertinentiis contra omnes homines [qui de stirpe suo exierint.]

[Here the fine, especially if it is of early date, will end; but in cases of subinfeudation, where rent or services are reserved, the following forms occur after 6.]

---

[1] Idem, or prælictus, or præfatus, or memoratus.

13.—Habend[a] et tenend[a] eidem A B, de prædicto C D, et heredibus suis imperpetuum (*or* tota vita sua).

14.—Reddendo inde per ann. (tota vita ipsius) ....*s.* ad terminos scilicet medietatem ad festum S⁻ᶜⁱ .... et aliam [alteram] ad festum Sᶜⁱ .... pro omnibus serviciis consuetudinibus et exactionibus ad prædictum C D, et heredes ejus pertinentibus.

15.—Et faciendo inde Capitalibus dominis de feodo (*or*, feodi illius) pro prædicto C D, omnia alia servicia quæ ad illa tenementa, etc., pertinent [salvo forinseco servitio].

16.—If the grant was for the tenant's life only, this occurs :—Et post decessum ipsius A B, prædicta tenementa cum pertinenciis integre revertentur ad prædictum C D, quieta de heredibus ipsius A B, tenenda de capitalibus dominis feodi illius per servitia quæ ad illa pertinent.

17.—If the rent was reserved during the grantor's life only, then this :—Et post decessum ipsius C D, prædictus A B, et heredes ejus erunt quieti de solutione prædicti redditus imperpetuum.

18.—If the fine is to entail the property the habendum will run thus :—Habendum et tenendum { eisdem A B, et E F, { eidem A B, et heredibus { et hered..de corporibus eorum inter eos. { quos idem A B, de corpore E F, uxoris ejus [legitime] procreaverit.

Form of Charter.

Univ'sis Francis et Anglis. { *Noverint* } { *Sciant* } *præsentes et futuri quod* ego, A B., etc.

Notum sit omnibus { Ecclesiæ fidelibus quod } { filiis Sanctæ Dei Ecclesiæ quod } ego, A. B.

Or :—

Omnibus Christi fidelibus ad quos hoc præsens Scriptum pervenerit A B, de, &c. Salutem in Domino Sempiterno. Noveritis me remississe, etc., or *Sciatis me præfatum, A B,* tam [pro naturali amore et affectione] [in parte complementum quarumdam conventionis concessionis et agreamenti contentarum specificatarum et declaratarum in quibusdam indenturis gerentibus data,

&c.] quam pro diversis aliis bonis et rationabilibus causis et consideraconibus me [ad hoc] specialiter movens.

Dono et concedo et per præsentem cartam confirmavi.

[Dedisse] concessisse feofasse liberasse et $^{hoc}_{hac}$ præsenti scripto meo $^{}$ confirmasse et per præsentes concedere feoffare Carta mea liberare et confirmare.

C D [filio meo].

[Omnes illas] sex $^{pecias}_{acras}$ terræ arabilis et unam acram prati continentes per estimacionem....

Sive plus, sive minus

Scituatas jacentes et existentes in parochia de G H.

Inter regiam [or altam] viam ducentem de . . . versus

Et ab antiquo acceptatas et reputatas ut pertinentes ad quoddam messuagium, etc., vel cum eodem occupatas et gavisas,

Quas quidem pecias terræ, etc., ego præfatus A B, nuper habui mihi et heredibus meis de dono concessione et charta Y Z, prout per chartam suam (dated, &c.) plenius liquet et apparet,

Habendum, tenendum, et gaudendum omnes illas, etc.

Et cetera omnia præmissa cum omnibus et singulis eorum et cujuslibet eorum pertinentiis.

Prædicto C D, heredibus et assignatis suis ad proprium opus et usum præfati C D, heredum et assignatorum suorum imperpetuum, *or*

Prædicto C D, et heredibus suis de capitalibus dominis feodi illius per servicia inde prius debita et de jure consueta imperpetuum per præsentes.

Warranty.—*Et ego vero prefatus A B, et heredes mei* prædictas sex acras, et cetera omnia et singula præmissa quæcumque cum suis et eorum cujuslibet pertinentibus præfato C D, heredibus et assignatis suis ad opera usus et intentiones prædicta *contra*[1] [*nos et heredes nostros*] *warrantizabimus* et imperpetuum defendemus per præsentes.

In cujus rei testimonium[2] huic præsenti cartæ meæ sigillum meum apposui. Dat' . . . die mensis . . . a° regni Regis . . .

---

[1] Absolute warranty would be "contra omnes gentes."

[2] Occasionally if parts were exchanged [una parte hujus præsentis scriptu indentati apud me, præfatum A B, remanente prædictus C D sigillum suum apposuit, alteri vero parti ego præfatus A B, sigillum meum apposui].

The memorandum endorsed as to the formal delivery of seizin usually ran thus :—

Status [possessio] et seisina [etiam capta et] deliberata fuerunt secundam formam et effectum præsentium die et anno infrascriptis in præsencia horum quorum nomina infra scribuntur.

G. H., J. K., L. M.,

Et mei N. O. notarii Publici.

APPENDIX II.

Registrar General's Documents.

Registers and records deposited in the custody of the Registrar General, General Register Office, Somerset House, London, W.C., where searches and certificates are granted between the hours of 10 and 4 daily (except on Sundays, Christmas Day, and Good Friday), on payment of the statutory fees :—

1. Registers of Births registered in England and Wales on and after 1st July, 1837.

2. Registers of Deaths registered in England and Wales on and after 1st July, 1837.

3. Registers of Marriages registered in England and Wales on and after 1st July, 1837, after solemnization in churches of the Established Church, in registered Roman Catholic and Dissenting places of worship, and in District Register Offices; also of Quakers and of Jews.

4. Registers of Births and Deaths at Sea registered since 1st July, 1837.

[The above Registers, Nos. 1, 2, 3, and 4, are made and kept pursuant to the Act 6 and 7 Will. IV. cap. 86.]

[The General Indexes of Births, Deaths, and Marriages, are completed about nine to twelve months after the date of the registration; but searches may be made in the registers not indexed if the *locality and date* can be accurately stated by the applicant.]

5. Non-parochial Registers of Baptisms or Births, Burials or Deaths, and (in a few instances) of Marriages, being the registers or records kept by various bodies and congregations of Nonconformists prior to the general system of registration commenced in 1837 ;—com-

prising, amongst others, the registers kept at *Dr. Williams's Library* from 1742, *Bunhill Fields Burial Ground* from 1713, the registers of *French Protestant and other Foreign Churches* in England, the registers, &c., of the *Society of Friends*, &c.

[By the Acts 3 and 4 Vict. cap. 92, and 21 Vict. cap. 25, *extracts from these registers, stamped with the seal of the General Register Office, are receivable in evidence in all civil cases.* When searches are required to be made, the description of register and the locality or name of the chapel should be given.]

6. Registers and Records of Baptisms and Marriages performed at the Fleet and King's Bench Prisons, at May Fair, at the Mint in Southwark, and elsewhere, between the years 1674 and 1754. These registers and records were transferred from the Registry of the Bishop of London to the custody of the Registrar General under the provisions of 3 and 4 Vict. cap. 92, sec. 20.

7. Registers of Births, Baptisms, Marriages, and Burials received from other places in England and Wales, comprising certain non-parochial registers received at the General Register Office since the passing of the Act 21 Vict. cap. 25, and certain registers received from the British Lying-in Hospital.

8. Registers of Births, Baptisms, Marriages, Deaths, and Burials of British Subjects in Foreign Countries, kept by British consuls and chaplains during the years 1809-1819 both inclusive. This volume consists of returns made by British consuls to the Foreign Office prior to the passing of the Act 12 and 13 Vict. cap. 68.

9. Registers of Marriages of British Subjects in Foreign Countries, solemnized by British consuls since July, 1849, under the provisions of the Act 12 and 13 Vict. cap. 68.

10. Registers of Births and Deaths of British Subjects in Foreign Countries, kept by British consuls since 7th November, 1849, and at British Legations since 19th July, 1859, in accordance with in-structional circulars issued from the Foreign Office on those dates.

11. Registers of Births, Baptisms, Marriages, Deaths, and Burials of British Subjects in Foreign Countries since the year 1784. These comprise certain original registers kept by British consuls and chaplains, and certain certificates, which have from time to time been received from the Foreign Office and other sources.

12. Registers and Certificates of Births, Marriages, and Deaths of British Subjects in Foreign Countries and in British Colonies and Possessions, kept and made by various British, Foreign, and Colonial authorities since 1801, and received through the Foreign and Colonial Offices.

13. Registers of Marriages in India solemnized since 1st January, 1852, in the presence of registrars, pursuant to 14 and 15 Vict. cap. 40. [Marriages solemnized by clergymen of the Church of England are not included in these returns.]

14. Duplicate Registers of Marriages performed by Chaplains in

the Bengal Presidency since July, 1864, and in Bombay in 1867 and 1868. (Indian Marriage Act, No 5, 1865.)

15. Registers of Marriages of British Subjects in the Ionian Islands, solemnized between the years 1861 and 1864, under the provisions of 23 and 24 Vict. cap. 86. This Act was repealed by 27 and 28 Vict. cap. 77, passed on 29th July, 1864, on the relinquishment of the Protectorate by Great Britain.

16. Registers of Births, Baptisms, Marriages, Deaths, and Burials, between the years 1818 and 1864, selected out of the Ionian Islands; papers received into the Public Record Office from the Colonial Office in July and August, 1864, upon the relinquishment of the Protectorate by Great Britain.

17. Registers of Baptisms, Marriages, and Burials by Army Chaplains in the Ionian Islands, between the years 1816 and 1864, transmitted to the General Register Office by the Secretary of State for War. These are original registers.

18. Registers of Baptisms, Marriages, and Burials of Military and Civilians, kept by army chaplains at the military stations at home and abroad from 1812-1870. These consist of returns received at the War Office.

19. Records and Returns of Buildings certified as places of meeting for religious worship, under the provisions of 15 and 16 Vict., cap. 36, and 18 and 19 Vict., cap. 81.

20. Calendars of the grants of Probate and Letters of Administration made in the Principal Registry and in the several District Registries of Her Majesty's Court of Probate, from the 11th January, 1858, pursuant to the Act 20 and 21 Vict. cap. 77.

APPENDIX III.

List of London Cemeteries.

Abney Park Cemetery, High Street, Stoke Newington.

Brompton Cemetery, Fulham Road, S.W.

City of London. Office, Sewers Office, Guildhall, E.C. ; Cemetery, Little Ilford.

City of London and Tower Hamlets, South Grove, Mile End Road.

Finchley. See Islington, St. Marylebone, and St. Pancras.

General Cemetery Co., Kensal Green. Office, 95, Great Russell Street, W.C.

East Metropolitan Cemetery, East Ham Park.

Great Northern London Cemetery, New Southgate, N.   Office, 22, Great Winchester Street, E.C.

Highgate. See London.

Islington Burial Board. Vestry Offices, Upper Street, Islington; Cemetery, East Finchley, N.

Jews' Burial Ground, 29, Alderney Road, E.

Jews' Burial Ground, Brady Street, E.

Jews' Burial Ground, Mile End Road, E.

Jews' Burial Ground, Queen's Elm, Fulham Road.

Jews' Cemeteries (United Synagogue), Willesden and West Ham.

Jews' Cemetery, Bancroft Road, E.

Kensington Burial Board. Vestry Hall, High Street, Kensington; Cemetery, Hanwell, W.

Lambeth Burial Board. Vestry Hall, Kennington Road; Cemetery, Lower Tooting, S.W.

London Cemetery. Office, 29, New Bridge Street, E.C.; Cemetery, Nunhead, S.E.

London Necropolis and Nat. Mausoleum. Office, Lancaster Place, W.C.; Cemetery, Woking.

Norwood. See South Metropolitan.

Nunhead. See London.

Paddington Burial Board. Vestry Hall, Harrow Road, W.; Cemetery, Willesden.

Roman Catholic (St. Mary's), Kensal Green, W.

St. George, Hanover Square. Burial Board, 9, Commercial Road, Pimlico; Cemetery, Hanwell, W.

St. Leonard's, Shoreditch, Burial Board, Town Hall Street, E.C.

St. Mary, Newington, Burial Board, Vestry Hall, Walworth Road, S.E.

St. Marylebone Burial Board, Court House, Marylebone Lane; Cemetery, East End, Finchley, N.

St. Pancras Burial Board, Vestry Hall, Pancras Road; Cemetery, East Finchley, N.

St. Saviour's, Southwark, Burial Board, Church Street, Boro' Market; Cemetery, Woking, Surrey.

South Metropolitan. Office, 13, New Bridge Street, E.C.; Cemetery, Lower Norwood, S.E.

Westminster (St. Anne, Soho) Burial Board, 17, Carlisle Street, Soho; Cemetery, Woking, Surrey.

Westminster (St. Margaret's and St. John's) Burial Board, 7, Westminster Chambers; Cemetery, Woking.

APPENDIX IV.

Classified List of the Master of the Rolls Series.

CHRONICLES AND HISTORIES.

ENGLAND.

26. Descriptive Catalogue of Manuscripts relating to the History of Great Britain and Ireland. Vol. i. (in Two Parts), Anterior to the Norman Invasion. Vol. ii., 1066-1200. Vol. iii., 1200-1327. By Sir Thomas Duffus Hardy, D.C.L. 1862-1871.

23. The Anglo-Saxon Chronicle, according to the several Original Authorities. Vol. i., Original Texts. Vol. ii., Translation. Edited and translated by Benjamin Thorpe, Esq. 1861.

3. Lives of Edward the Confessor. I., La Estoire de Seint Aedward le Rei. II., Vita Beati Edvardi Regis et Confessoris. III., Vita Æduuardi Regis qui apud Westmonasterium requiescit. Edited by Henry Richards Luard, M.A. 1858.

30. Ricardi de Cirencestria Speculum Historiale de Gestis Regum Angliæ. Vol. i., 447-871. Vol. ii., 872-1066. Edited by John E. B. Mayor, M.A. 1863-1869.

44. Matthæi Parisiensis Historia Anglorum, sive, ut vulgo dicitur, Historia Minor. Vols. i., ii., and iii., 1067-1253. Edited by Sir Frederic Madden, K.H. 1866-1869.

57. Matthæi Parisiensis, Monachi Sancti Albani, Chronica Majora. Vol. i., The Creation to A.D. 1066. Vol. ii., A.D. 1067 to A.D. 1216. Vol. iii., A.D. 1216 to A.D. 1239. Vol. iv., A.D. 1240 to A.D. 1247. Vol. v., A.D. 1248 to A.D. 1259. Vol. vi., Additamenta. Vol. vii., Index. Edited by Henry Richards Luard, D.D. 1872-1884.

36. Annales Monastici. Vol. i., Annales de Margan, 1066-1232; Annales de Theokesberia, 1066-1263; Annales de Burton, 1004-1263. Vol. ii., Annales Monasterii de Wintonia, 519-1277; Annales Monasterii de Waverleia, 1-1291. Vol. iii., Annales Prioratus de Dunstaplia, 1-1297; Annales Monasterii de Bermundesoia, 1042-1432. Vol. iv., Annales Monasterii de Osenoia, 1016-1347; Chronicon vulgo dictum Chronicon Thomæ Wykes, 1066-1289; Annales Prioratus de Wigornia, 1-1377. Vol. v., Index and Glossary. Edited by Henry Richards Luard, M.A. 1864-1869.

52. Willelmi Malmesbiriensis Monachi de Gestis Pontificum Anglorum Libri Quinque. Edited, from William of Malmesbury's Autograph MS., by N. E. S. A. Hamilton. 1870. [Ends 1122.]

68. Radulfi de Diceto Decani Lundoniensis Opera Historica. The Historical Works of Master Ralph de Diceto, Dean of London. Vols. i. and ii. Edited, from the Original Manuscripts, by

William Stubbs, M.A. 1876. [Part ends 1147, and other part 1201.]

74. Henrici Archidiaconi Huntendunensis Historia Anglorum. The History of the English, by Henry, Archdeacon of Huntingdon, from A.D. 55 to A.D. 1154, in Eight Books. Edited by Thomas Arnold, Esq., M.A. 1879. [Ends 1154.]

51. Chronica Magistri Rogeri de Houedene. Vols. i., ii., iii., and iv. Edited by William Stubbs, M.A. 1868-1871. [Valuable from 1192 to 1201.]

75. The Historical Works of Symeon of Durham. Vols. i. and ii. Edited by Thomas Arnold, Esq., M.A., of Univ. College, Oxford. 1882-5. [Ends 1156.]

21. The Works of Giraldus Cambrensis. Vols. i., ii., iii., and iv. Edited by J. S. Brewer, M.A., Prof. of Engl. Liter., King's Coll., London. Vols. v., vi., and vii. Edited by the Rev. James F. Dimock, M.A., Rector of Barnburgh, Yorkshire. 1861-1877. [Ends 1187.]

82. Chronicles of the Reigns of Stephen, Henry II., and Richard I. Vols. i. and ii. 1884, 1885. Edited by Richard Howlett. [Ends 1198.]

49. Gesta Regis Henrici Secundi Benedicti Abbatis. Chronicle of the Reigns of Henry II. and Richard I., 1169-1192, known under the name of Benedict of Peterborough. Vols. i. and ii. Edited by William Stubbs, M.A. 1867.

73. Historical Works of Gervase of Canterbury. Vols. i. and ii. The Chronicle of the Reigns of Stephen, Henry II., and Richard I., by Gervase, the Monk of Canterbury. Edited by William Stubbs, D.D. 1879, 1880. [Ends 1199.]

38. Chronicles and Memorials of the Reign of Richard the First. Vol. i., Itinerarium Peregrinorum et Gesta Regis Ricardi. Vol. ii., Epistolae Cantuarienses; the Letters of the Prior and Convent of Christ Church, Canterbury; 1187 to 1199. Edited by William Stubbs, M.A. 1864-1865.

66. Radulphi de Coggeshall Chronicon Anglicanum. Edited by the Rev. Joseph Stevenson, M.A. 1875. [Ends 1200.]

27. Royal and other Historical Letters illustrative of the Reign of Henry III. Vol. i., 1216-1235. Vol. ii., 1236-1272. Selected and edited by the Rev. W. W. Shirley, D.D. 1862-1866.

28. Chronica Monasterii S. Albani.—1. Thomae Walsingham Historia Anglicana; Vol. i., 1272-1381: Vol. ii., 1381-1422. 2. Willelmi Rishanger Chronica et Annales, 1259-1307. 3. Johannis de Trokelowe et Henrici de Blaneforde Chronica et Annales, 1259-1296; 1307-1324; 1392-1406. 4. Gesta Abbatum Monasterii S. Albani, a Thoma Walsingham, regnante Ricardo Secundo, ejusdem Ecclesiae Praecentore compilata; Vol. i., 793-1290: Vol. ii., 1290-1349: Vol. iii., 1349-1411. 5. Johannis Amundesham, Monachi Monasterii S. Albani, ut videtur, Annales; Vols. i. and ii.

6. Registra quorundam Abbatum Monasterii S. Albani, qui sæculo xv^mo floruere; Vol. i., Registrum Abbatiæ Johannis Whethamstede, Abbatis Monasterii Sancti Albani, iterum susceptæ; Roberto Blakeney, Capellano, quondam adscriptum : Vol. ii., Registra Johannis Whethamstede, Willelmi Albon, et Willelmi Walingforde, Abbatum Monasterii Sancti Albani, cum Appendice, continente quasdam Epistolas, a Johanne Whethamstede Conscriptas. 7. Ypodigma Neustriæ a Thoma Walsingham, quondam Monacho Monasterii S. Albani, conscriptum. Edited by Henry Thomas Riley, Esq., M.A. 1863-1876.

58. Memoriale Fratris Walteri de Coventria, the Historical Collections of Walter of Coventry. Vols. i. and ii. Edited, from the MS. in the Library of Corpus Christi College, Cambridge, by William Stubbs, M.A. 1872-1873. [Ends 1225.]

84. Chronica Rogeri de Wendover, sive Flores Historiarum. Vol. i. Edited by Henry Gay Hewlett. [Ends 1235.]

42. Le Livere de Reis de Brittanie e Le Livere de Reis de Engletere. Edited by John Glover, M.A. 1865.

13. Chronica Johannis de Oxenedes. Edited by Sir Henry Ellis, K.H. 1859.

16. Bartholomæi de Cotton, Monachi Norwicensis, Historia Anglicana ; 449-1298 : necnon ejusdem Liber de Archiepiscopis et Episcopis Angliæ. Edited by Henry Richards Luard, M.A. 1859. [Ends 1298.]

76. Chronicles of the Reigns of Edward I. and Edward II. Vols. i. and ii. Edited by William Stubbs, D.D. 1882, 1883. [Ends 1272.]

47. The Chronicle of Pierre de Langtoft, in French verse, from the earliest period to the death of Edward I. Vols. i. and ii. Edited by Thomas Wright, Esq., M.A. 1866-1868.

9. Eulogium (Historiarum sive Temporis) : Chronicon ab Orbe condito usque ad Annum Domini 1366 ; a Monacho quodam Malmesbiriensi exaratum. Vols. i., ii., and iii. Edited by F. S. Haydon, B.A. 1858-1863. [Ends 1366.]

41. Polychronicon Ranulphi Higden, with Trevisa's Translation. Vols. i. and ii. Edited by Churchill Babington, B.D., Senior Fellow of St. John's College, Cambridge. Vols. iii., iv., v., vi., vii., viii., and ix. Edited by the Rev. Joseph Rawson Lumby, D.D. 1865-1886. [Ends 1377.]

64. Chronicon Angliæ, ab Anno Domini 1328 usque ad Annum 1388, Auctore Monacho quodam Sancti Albani. Edited by Edward Maunde Thompson. 1874. [Ends 1388.]

18. A Collection of Royal and Historical Letters during the Reign of Henry IV. 1399-1404. Edited by the Rev. F. C. Hingeston, M.A. 1860. [Ends 1404.]

1. The Chronicle of England, by John Capgrave. Edited by the Rev. F. C. Hingeston, M.A. 1858. [Ends 1417.]

29. Chronicon Abbatiæ Eveshamensis, Auctoribus Dominico Priore Eveshamiæ et Thoma de Marleberge Abbate, a Fundatione ad Annum 1213, una cum Continuatione ad Annum 1418. Edited by the Rev. W. D. Macray. 1863. [Ends 1418.]

11. Memorials of Henry the Fifth. I., Vita Henrici Quinti, Roberto Redmanno auctore. II., Versus Rhythmici in laudem Regis Henrici Quinti. III., Elmhami Liber Metricus de Henrico V. Edited by Charles A. Cole. 1858. [Ends 1422.]

39. Recueil des Croniques et anchiennes Istories de la Grant Bretaigne a present nommée Engleterre, par Jehan de Waurin. Vol. i. Albina to 688. Vol. ii., 1399-1422. Vol. iii., 1422-1431. Edited by Sir William Hardy, F.S.A. 1864-1879. Vol. iv., 1431-1443. Edited by Sir William Hardy, F.S.A., and Edward L. C. P. Hardy, F.S.A. 1884. [Ends 1443.]

40. A Collection of the Chronicles and Ancient Histories of Great Britain, now called England, by John de Wavrin. Albina to 688. Translation of, by Sir W. Hardy. 1864.

14. A Collection of Political Poems and Songs relating to English History, from the Accession of Edward III. to the Reign of Henry VIII. Vols. i. and ii. Edited by Thomas Wright, M.A. 1859-1861.

32. Narratives of the Expulsion of the English from Normandy, 1449-1450.—Robertus Blondelli de Reductione Normanniæ: Le Recouvrement de Normandie, par Berry, Hérault du Roy: Conferences between the Ambassadors of France and England. Edited, from MSS. in the Imperial Library at Paris, by the Rev. Joseph Stevenson, M.A. 1863. [Ends 1450.]

56. Memorials of the Reign of Henry VI.:—Official Correspondence of Thomas Bekynton, Secretary to Henry VI., and Bishop of Bath and Wells. Edited from a MS. in the Archiepiscopal Library at Lambeth, with an Appendix of Illustrative Documents, by the Rev. George Williams, B.D. Vols. i. and ii. 1872. [Ends 1461.]

22. Letters and Papers illustrative of the Wars of the English during the Reign of Henry the Sixth, King of England. Vol. i. and Vol. ii. (in Two Parts.) Edited by the Rev. Joseph Stevenson, M.A. 1861-1864. [Ends 1461.]

7. Johannis Capgrave Liber de Illustribus Henricis. Edited by the Rev. F. C. Hingeston, M.A. 1858. [Ends 1461.]

24. Letters and Papers illustrative of the Reigns of Richard III. and Henry VII. Vols. i. and ii. Edited by James Gairdner. 1861-1863. [Ends 1509.]

60. Materials for a History of the Reign of Henry VII., from original Documents preserved in the Public Record Office. Vols. i. and ii. Edited by the Rev. William Campbell, M.A. 1873-1877. [Ends 1509.]

10. Memorials of Henry the Seventh: Bernardi Andreæ Tholosatis,

Vita Regis Henrici Septimi; necnon alia quædam ad eundem Regem spectantia. Edited by James Gairdner. 1858. [Ends 1509.]

IRELAND.

48. The War of the Gaedhil with the Gaill, or the Invasions of Ireland by the Danes and other Norsemen. Edited, with a Translation, by James Henthorn Todd, D.D. 1867.

54. The Annals of Loch Cé. A Chronicle of Irish Affairs, from 1041 to 1590. Vols. i. and ii. Edited, with a Translation, by William Maunsell Hennessy. 1871.

46. Chronicon Scotorum: a Chronicle of Irish Affairs, from the Earliest Times to 1135; and Supplement, containing the Events from 1141 to 1150. Edited, with Translation, by William Maunsell Hennessy. 1866.

53. Historic and Municipal Documents of Ireland, from the Archives of the City of Dublin, &c. 1172-1320. Edited by John T. Gilbert. 1870.

69. Roll of the Proceedings of the King's Council in Ireland, for a Portion of the 16th Year of the Reign of Richard II. 1392-93. Edited by the Rev. James Graves, A.B. 1887.

SCOTLAND.

6. The Buik of the Croniclis of Scotland; or, A Metrical Version of the History of Hector Boece; by William Stewart. Vols. i., ii., and iii. Edited by W. B. Turnbull. 1858.

WALES.

20. Annales Cambriæ. 447-1288. Edited by the Rev. John Williams ab Ithel, M.A. 1860.

17. Brut y Tywysogion; or, the Chronicle of the Princes of Wales. 681-1282. Edited by the Rev. John Williams ab Ithel, M.A. 1860.

MONASTIC CHARTULARIES AND CHRONICLES.

[As to St. Alban's, see Historical Chronicles, 28, Nos. 5 and 7. As to Westminster (Charters), see Historical Chronicles, 30 (Richard de Cirencestria.)]

2. Chronicon Monasterii de Abingdon. Vols. i. and ii. Edited by the Rev. Joseph Stevenson, M.A. 1858.

8. Historia Monasterii S. Augustini Cantuariensis, by Thomas of Elmham, formerly Monk and Treasurer of that Foundation. Edited by Charles Hardwick, M.A. 1858.

85. The Registers of the Monastery of Christ Church, Canterbury. Edited by Joseph Brigstocke Sheppard, LL.D. Vol. i.

80. Chartularies of St. Mary's Abbey, Dublin, &c., preserved in the

Bodleian Library and British Museum. Edited by John Thomas Gilbert. Vols. i. and ii. 1884 and 1885.

33. Historia et Cartularium Monasterii S. Petri Gloucestriæ. Vols. i., ii., and iii. Edited by W. H. Hart. 1863-1867.

45. Liber Monasterii de Hyda: a Chronicle and Chartulary of Hyde Abbey, Winchester, 455-1023. Edited, from a Manuscript in the Library of the Earl of Macclesfield, by Edward Edwards. 1866.

72. Registrum Malmesburiense : the Register of Malmesbury Abbey; preserved in the Public Record Office. Vols. i. and ii. Edited by J. S. Brewer, M.A., and Charles Trice Martin, Esq., B.A. 1879, 1880.

43. Chronica Monasterii de Melsa ab Anno 1150 usque ad Annum 1406. Vols. i., ii., and iii. Edited by Edward Augustus Bond. 1866-1868.

83. Chronicle of the Ancient Abbey of Ramsey, from the Chartulary of that Abbey in the Public Record Office. 1886. Edited by the Rev. William Dunn Macray, M.A.

79. Chartulary of the Ancient Benedictine Abbey of Ramsey, from the MS. in the Public Record Office. Vols. i. and ii. 1884, 1886. Edited by William Henry Hart and the Rev. Ponsonby Annesley Lyons.

4. Monumenta Franciscana. Vol. i., Thomas de Eccleston de Adventu Fratrum Minorum in Angliam. Adæ de Marisco Epistolæ. Registrum Fratrum Minorum Londoniæ. Edited by J. S. Brewer, M.A. Vol. ii., De Adventu Minorum ; re-edited, with additions. Chronicle of the Grey Friars. The ancient English version of the Rule of St. Francis. Abbreviatio Statutorum, 1451, &c. Edited by Richard Howlett. 1858, 1882.

67. Materials for the History of Thomas Becket, Archbishop of Canterbury. Vols. i., ii., iii., iv., v., and vi. Edited by the Rev. James Craigie Robertson, M.A. 1875-1883. Vol. vii. Edited by Joseph Brigstocke Sheppard, LL.D. 1885.

65. Thómas Saga Erkibyskups. A Life of Archbishop Thomas Becket, in Icelandic. Vols. i. and ii. Edited, with English Translation, Notes, and Glossary, by M. Eiríkr Magnússon, M.A. 1875-1884.

37. Magna Vita S. Hugonis Episcopi Lincolniensis. From MSS. in the Bodleian Library, Oxford, and the Imperial Library, Paris. Edited by the Rev. James F. Dimock, M.A. 1864.

25. Letters of Bishop Grosseteste, illustrative of the Social Condition of his Time. Edited by Henry Richards Luard, M.A. 1861.

63. Memorials of Saint Dunstan, Archbishop of Canterbury. Edited, from various MSS., by William Stubbs, M.A. 1874.

71. The Historians of the Church of York, and its Archbishops. Vols. i. and ii. Edited by James Raine, M.A. 1879-1886.

61. Historical Papers and Letters from the Northern Registers. Edited by James Raine, M.A. 1873.

62. Registrum Palatinum Dunelmense. The Register of Richard de

Kellawe, Lord Palatine and Bishop of Durham; 1311-1316. Vols. i., ii., iii., and iv. Edited by Sir Thomas Duffus Hardy, D.C.L. 1873-1878.

77. Registrum Epistolarum Fratris Johannis Peckham Archiepiscopi Cantuariensis. Vols. i. ii., and iii. Edited by Charles Trice Martin, Esq., B.A. 1882-1886.

78. Register of S. Osmund. Edited by the Rev. W. H. Rich Jones, M.A. Vols. i. and ii. 1883, 1884.

81. Eadmeri Historia Novorum in Anglia, et Opuscula duo de Vita Sancti Anselmi et quibusdam Miraculis ejus. 1884. Edited by the Rev. Martin Rule, M.A.

19. The Repressor of over much Blaming of the Clergy. By Reginald Pecock. Vols. i. and ii. Edited by Churchill Babington, B.D. 1860.

5. Fasciculi Zizaniorum Magistri Johannis Wyclif cum Tritico. Ascribed to Thomas Netter, of Walden, Provincial of the Carmelite Order in England, and Confessor to King Henry the Fifth. Edited by Rev. W. Shirley, M.A. 1858.

50. Munimenta Academica, or, Documents illustrative of Academical Life and Studies at Oxford (in Two Parts.) Edited by the Rev. Henry Anstey, M.A. 1868.

MUNICIPAL.

12. Munimenta Gildhallæ Londoniensis; Liber Albus, Liber Custumarum, et Liber Horn, in archivis Gildhallæ asservati. Vol. i., Liber Albus. Vol. ii. (in Two Parts), Liber Custumarum. Vol. iii., Translation of the Anglo-Norman Passages in Liber Albus, Glossaries, Appendices, and Index. Edited by Henry Thomas Riley, M.A. 1859-1862.

PHILOSOPHY. MEDICINE. SCIENCE.

35. Leechdoms, Wortcunning, and Starcraft of Early England; being a Collection of Documents illustrating the History of Science in this Country before the Norman Conquest. Vols. i., ii., and iii. Collected and edited by the Rev. T. Oswald Cockayne, M.A. 1864-1866.

15. The "Opus Tertium," "Opus Minus," &c., of Roger Bacon. Edited by J. S. Brewer, M.A. 1859.

34. Alexandri Neckam de Naturis Rerum, libri duo; with Neckam's Poem, De Laudibus Divinæ Sapientiæ. Edited by Thomas Wright, M.A. 1863.

SATIRICAL POETRY.

59. The Anglo-Latin Satirical Poets and Epigrammatists of the Twelfth Century. Vols. i. and ii. Collected and edited by Thomas Wright, M.A. 1872.

14. Political Poems and Songs, Edward III. to Henry VIII. Thomas Wright. Vols. i. and ii. 1859-1861.

LAW.

31. Year Books of the Reign of Edward the First. Years 20-21, 21-22, 30-31, 32-33, and 33-35 Edw. I; and 11-12 Edw. III., edited and translated by Alfred John Horwood. Years 12-13, 13-14 Edward III., edited and translated by Luke Owen Pike, Esq., M.A. 1863-1866.

70. Henrici de Bracton de Legibus et Consuetudinibus Angliæ, Libri Quinque in Varios Tractatus Distincti. Ad Diversorum et Vetustissimorum Codicum Collationem Typis Vulgati. Vols. i., ii., iii., iv., v., and vi. Edited by Sir Travers Twiss, Q.C., D.C.L. 1878-1883.

55. Monumenta Juridica. The Black Book of the Admiralty, with Appendices. Vols. i., ii., iii., and iv. Edited by Sir Travers Twiss, Q.C., D.C.L. 1871-1876.

APPENDIX V.

𝔇𝔬𝔪𝔢𝔰𝔱𝔦𝔠 𝔖𝔱𝔞𝔱𝔢 𝔓𝔞𝔭𝔢𝔯𝔰.

EDWARD VI. TO JAMES I.

By R. Lemon.

1547—80	. . .	vol. 1
1581—90	. . .	2
1547-1565, see Addenda		6
1566—1579	,,	7
1580—1625	,,	12
1603—1625	,,	11

Mrs. Everett Green.

1591—4	. . .	3
1595—7	. . .	4
1598—1601	. .	5
1601—1603	. .	6
[1547—1565 Addenda]		6
[1566—1579 Addenda]		7

EDWARD VI. TO JAMES I. (continued).

1603—1610	. .	vol. 8
1611—1618	. .	9
1619—1623	. .	10
1623—1625	. .	11
[1603—1625 Addenda]		11
[1580—1625 Addenda]		12

CHARLES I.

J. Bruce.

1625—6	. . .	1
1627—8	. . .	2
1628—9	. . .	3
1629—31	. . .	4
1631—3	. . .	5
1633—4	. . .	6

CHARLES I. *(continued.)*

1634—5	.	.	.	vol. 7
1635		.	.	8
1635—6		.	.	9
1636—7		.	.	10
1637		.	.	11
1637—8	.	.	.	12

Bruce and Hamilton.

1638—9	.	.	.	13

W. D. Hamilton.

1639		.	.	14
1639—40		.	.	15
1640		.	.	16
1640—1	.	.	.	17
1641—3, in the press				18

[Still in progress.]

COMMONWEALTH.

Mrs. Everett Green.

1649	.	.	.	1
1650	.	.	.	2
1651	.	.	.	3
1651—2	.	.	.	4
1652—3	.	.	.	5
1653—4	.	.	.	6
1654	.	.	.	7
1655	.	.	.	8
1655—6	.	.	.	9
1656—7	.	.	.	10
1657—8	.	.	.	11
1658—9	.	.	.	12
1659—60	.	.	.	13

Vol. 14 in the press.

CHARLES II.

Mrs. Everett Green.

1660—1	.	.	.	1
1661—2	.	.	.	2
1663—4	.	.	.	3
1664—5	.	.	.	4

CHARLES II. *(continued.)*

1665—6	.	.	.	vol. 5
1666—7	.	.	.	6
1667		.	.	7

Home Office Papers.

GEORGE III.

J. Redington.

1760—5	.	.	.	1
1766—9	.	.	.	2

R. A. Roberts.

1770—2	.		3

Foreign & Domestic Series.

HENRY VIII.

J. S. Brewer.

1509—14	.	vol. 1
1515—18	.	2 (2 Pts.)
1519—23	.	3 (2 Pts.)
[Introduction]		4, Pt.
1524—6	.	4, Pt. 1
1526—8	.	4, Pt. 2
1529—30	.	4, Pt. 3

J. Gairdner.

1531—2	.	.	.	vol. 5
1533	.	.	.	6
1534	.	.	.	7
1535 (to January)	.			8
1535 (to December)	.			9

Vol. 10 in the press.

Treasury Papers.[1]

J. Redington.

1557—1696	.	.	1
1697—1702	.	.	2
1702—1707	.	.	3
1708—1714	.	.	4
1714—1719	.	.	5

[1] Give pensions, grants, remissions, &c.

Colonial Series.

1574—1668.

W. Noel Sainsbury.

1574—1660.	Vol 1.	America and West Indies.
1513—1616.	2.	East Indies, China, and Japan.
1617—1621.	3.	East Indies, China, and Japan.
1622—1624.	4.	East Indies, China, and Japan.
1661—1668.	5.	America and West Indies.
1625—1629.	6.	East Indies.

There are also the following Scottish and Irish papers :—

State Papers, Scottish Series.

HENRY VIII.—ELIZABETH.

M. J. Thorpe.

1509—1589	. .	vol. 1
1589—1603	. .	2
1543—1592	.	Appendix

Documents, Ireland.

1171—1307.

H. S. Sweetman, continued by G. F. Handcock.

1171—1251	. .	vol. 1
1252—1284	. .	2
1285—1292	. .	3
1293—1301	. .	4
1302—1307	. .	5

State Papers (Ireland)

Of the reigns of HENRY VIII., EDWARD VI., MARY, and ELIZABETH.

H. C. Hamilton.

1509—1573	.	vol. 1
1574—1585	.	2
1586—1588	.	3
1588—1592	.	4

State Papers (Ireland.)

JAMES I.

Rev. C. W. Russell and J. P. Prendergast.

1603—1606	. .	vol. 1
1606—1608	. .	2
1608—1610	. .	3
1611—1614	. .	4
1615—1625	. .	5

Carew Papers (at the Lambeth Library.)

1515—1624.

1515—1574	. .	1
1575—1588	. .	2
1589—1600	. .	3
1601—1603	. .	4
Book of Howth Miscellaneous		5
1603—1624	. .	6

Foreign Series.

EDWARD VI.

W. B. Turnbull.
1547—1553

MARY.

W. B. Turnbull.
1553—1558

ELIZABETH.

Rev. J. Stevenson, M.A.

1558—1559	.	. vol.	1
1559—1560	.	.	2
1560—1561	.	.	3
1561—1562	.	.	4
1562 .	.	.	5
1563 .	.	.	6
1564—1565	.	.	7

A. J. Crosby, M.A.

1566—1568	.	.	8
1569—1571	.	.	9
1572—1574	.	.	10
1575—1577	.	.	11

𝔉oreign (𝔖panish, &c.)

HENRY VII.—HENRY VIII.

G. A. Bergenroth.

1485—1509	.	. 1
1509—1525	.	. 2

Supplement to vols. 1 and 2.

HENRY VII.—HENRY VIII. (*continued.*)

Don Pascual de Gayangos.

1525—1526	vol. 3,	Part 1
1527—1529	3,	2
1529—1530	4,	1
1531—1533	4,	2
1531—1533	4 continued	
1534—1536	5, Part 1	

𝔙enetian, &c.

1202—1558.

Rawdon Brown.

1202—1509	.	vol.	1
1509—1520	.		2
1520—1526	.		3
1527—1533	.		4
1534—1554	.		5
1555—1556	.		6, Part 1
1556—1557	.		6, 2
1557—1558	.		6, 3

APPENDIX VI.

𝔥istorical 𝔐𝔖𝔖. 𝔠ommission.

The following are Lists of the Reports issued by the Historical MSS. Commission to date. Part I. contains Reports on muniment rooms, &c., of various corporations, colleges, towns, and parishes; and Part II. on MSS. belonging to private persons. The figures refer to the report and the page respectively :—

PART I.

Abingdon, Corporation, 1, p. 98; 2, p. 149
,, Christ's Hospital, 1, p. 98
Alwington, Parish, 5, p. 597

Ampleforth. St. Lawrence College, 2, p. 109
Axbridge, Corporation of, 3, p. 300
Barnstaple, Corporation, 9, p. 203

Berwick-on-Tweed, Town, 3, p. 308; Treasurer, 3, p. 309
Bishop's Castle, Corporation, 10th Rep., App. IV.
Bridgenorth, Corporation, 10th Rep., App. IV.
Bridgwater, Corporation, 1, p. 99; 3, p. 310
Bristol, Dean and Chapter, 1, p. 97
Cambridge, Corporation, 1, p. 99
,, Christ's, 1, p. 63
,, Clare, 2, p. 110
,, Corpus, 1, p. 64
,, Downing, 3, p. 320
,, Emanuel, 4, p. 417
,, Gonville and Caius, 2, p. 116
,, Jesus, 2, p. 118
,, King's, 1, p. 67
,, Magdalen, 5, p. 481
,, Pembroke, 1, p. 67; 5, p. 484
,, Queen's, 1, p. 72
,, Regist. of University, 1, p. 73
,, St. Catherine's, 4, p. 421
,, St. John's, 1, p. 74
,, Sidney Sussex, 3, p. 327
,, Trinity, 1, p. 82
,, Trinity Hall, 2, p. 121
Canterbury, Dean & Chapter of, 5, p. 426; 8, p. 315; 9, p. 72
Carisbrooke, 6, p. 499
Carlisle, Corporation of, 9, p. 197
Carlisle, Dean and Chapter, 2, p. 123
Chedder, Parish of, 3, p. 329
Chester, Corporation of, 8, p. 355
Chetham Library, Manchester, 2, p. 156
Cinque Ports, 4, p. 428
Coventry, Corporation, 1, p. 100
Dartmouth, Corporation, 5, p. 597
Dublin, See of, 10th Rep., App. V

Essex, County, 10th Rep., App. IV.
Eton College, 9, p. 349
Ewelore, Almshouse, 8, p. 624; 9, p. 216
Eye, Corporation, 10th Rep., App. IV.
Faversham, Corporation, 6, p. 500
Folkestone, Corporation, 5, p. 590
Fordwich, Corporation, 5, p. 606
Galway, 10th Rep., App. V.
Glastonbury, Town, 1, p. 102
Guernsey, Dr. Harkin's MSS., 2, p. 158
Hartland, Parish, 4, p. 428; 5, p. 571
Hunstanton, 9, p. 358
Hythe, Corporation, 4, p. 428
Hythe, St. Bartholomew's Hospital, 6, p. 511
Inner Temple Library, (Petyt MSS.), 2, p. 151
Ipswich, Corporation, 9, p. 222
Ireland, Jesuits in, 10th Rep., App. V.
Kendal, Corporation, 10th Rep., App. IV.
Kingston-on-Thames, Corporation, 3, p. 331
Lambeth, Palace, 6, p. 522
Laud, Archbishop, Visitation, 2, p. 108; 4, p. 124
Launceston, Corporation, 6, p. 524
Leicester, Corporation of, 8, p. 403
London, Catholic Chapter of, 5, p. 463
Lords, House of, 1, p. 1; 2, p. 106; 3, p. 1; 4, pp. 1-163; 5, pp. 1-120; 6, p. 1; 7, p. 1; 8, p. 101; 9, pt. 2, p. 1
Lords, House of, (1678-1688), 11th Rep., part 2
Lydd, Corporation, 5, p. 516
Lynn, King's, Corporation, 11th Rep., App. pt. 3
Manchester, Chetham Library, 2, p. 156

Mendlesham, Parish, 5, p. 590

Morpeth, Corporation, 6, p. 526

Norwich, Bishop's Registry, 1, p. 86

,, Corporation, 1, p. 102

,, Dean and Chapter, 1, p. 87

Nottingham, Corporation, 1, p. 105

Oscott, St. Mary's College, 1, p. 89 ; 2, p. 125

Ossory, See of, 10th Rep., App. v.

Oswestry, Corporation, 10th Rep., App.

Oxford, Balliol, 4, p. 442

,, Corpus, 2, p. 126

,, Exeter, 2, p. 127

,, Jesus, 2, p. 130

,, Lincoln, 2, p. 130

,, Magdalen, 4, p. 458; 8, p. 262

,, Merton, 6, p. 545

,, New, 2, p. 133

,, Oriel, 2, p. 136

,, Pembroke, 6, p. 549

,. Queen's, 2, p. 137; 4, p. 451; 6, p. 551

,, St. John's, 4, p. 465

,, Trinity, 2, p. 142

,, University Hall, 5, p. 477

,, Wadham, 5, p. 479

,. Worcester, 2, p. 143

Parkham, Parish of, 4, p. 468

Petersfield, Corporation, 10th Rep., App.

Physicians, Royal College of, 8, p. 226

Plymouth, Corporation, 9, p. 262 ; 10th Rep., App. iv.

Pontefract, Corporation of, 8, p. 269

Queen Anne's Bounty, 8, p. 632

Rochester, Bridge Warden, 9, p. 285

Rochester, Corporation, 9, p. 286

Romney, New, Corporation. 4, p. 439 ; 5, p. 533

,, Court Books, 6, p. 540

Rye, Corporation, 5, p. 488

Salisbury Cathedral, 1, p. 90

Sandwich, Corporation, 5, p. 568

Somerset, Consistory, 3, p. 333 ; 7, p. 623

Southampton, Corporation, 11th Rep., App. pt. 3

Southampton, God's House at, 6, p. 551

St. Alban's, Corporation, 5, p. 565

Stoneyhurst, College, 2. p. 143 ; 3, p. 334; 10th Rep., App. iv.

St. Paul's, Dean and Chapter, 9, p. 1

Stratford-on-Avon, Corporation, 9, p. 289

Totnes, Corporation, 3, p. 341

Trinity House, 8, p. 234

Ushaw, College, 1, p. 91

Waterford, Corporation of, 10th Rep., App. v.

Wells, Almshouses, 8, p. 638

,, Bishop's Registry, 1, p. 92

,, Cathedral, 10th Rep., App.

,, City, 3, p. 350

,, Corporation, 1, p. 106

,, Dean and Chapter, 1, p. 93 ; 3, p. 351

,, Vicar's Choral of, 3, p. 364

Wenlock, Corporation, 10th Rep., App. iv.

Westminster Abbey, 4, p. 171 ; 1, p. 94

Weymouth and Melcombe, Corporation Registers, 5, p. 575

Williams, Dr., Library

Wisbeach, Corporation, 9, p. 293

Woodchester, Monastery of Dominican Friars, 2, p. 146

Worcester, St. Andrew's, 8, p. 638

Wycombe, High, Borough of, 5, p. 554
Yarmouth, Great, Corporation, 9, p. 299
York, Corporation, 1, p. 108
,, Dean and Chapter, 1, p.97

York, Merchants Adventurers, 1, p. 110
Yorkshire Philosophical Society, 1, p. 110
Yorkshire, N. Riding, 9, p. 329
,, W. Riding, 9, p. 324

Part II.

Abergavenny, Marquis of, 10th Rep., App. VI.
Acland-Hood, Sir A., 7, p. 344
Ainslie, Miss, 2, p. 68
Almack, R., 1, p. 55
Antrobus, J. C., 2, p. 69
Arundell of Wardour, Lord, 2, p. 33
Ashburnham, Earl of, 8, part 3, p. 1
Bagot, Captain J. F., 10th Rep., App. pt. 4
Bagot, Lord, 4, p. 325
Baker, W. R., 2, p. 69
Balfour, B. R. T., 10th Rep., pt. 6
Bankes, R., 8, p. 208
Bath, Marquis of, 3, p. 180; 4, p. 227
Beaumont, W., 4, p. 368
Bedford, Duke of, 2, p. 1
Bedingfield, Sir H., 3, p. 237
Berington, C. M., 2, p. 72
Bouverie, P. Pleydell, 10th Rep., App. VI.
Bouverie-Pusey, S. E. E.,7, p. 681
Boycott, The Misses, 10th Rep., App. VI.
Bradford, Earl of, 2, p. 30
Braybrooke, Lord, 8, p. 277
Braye, Lord, 10th Rep., App. VI.
Bromley-Davenport, W., 2, pp. 78-80
Browne, Geo.,10th Rep., App. IV.
Brumell, F., 6, p. 538
Bunbury, Sir C., 3, p. 240
Bute, Marquis of, 3, p. 202
Calthorpe, Lord, 2, p. 39
Camoys, Lord, 2, p. 33

Carew, Colonel, 2, p. 74
Carew, Lieutenant-Colonel, 4, p. 368
Cathcart, Earl, 2, p. 24
Cawdor, Earl, 2, p. 31
Chichester, Earl of, 3, p. 221
Cholmondeley, R., 5, p. 333
Colchester, Lord, 4, p. 344
Collis, Mrs., 2, p. 76
Cooke, P. B. D., 6, 418
Cope, Sir W., 3, p. 242
Corbet, Rev. J. D., 10, p. 19
Corbet, R., 2, p. 77
Coventry, Earl of, 1, p. 34
Cowper, Countess, 2, p. 4
Dartmouth, Earl of, 2, p. 9
Dasent, Sir G. W., 6, p. 407
Davenport, W. Bromley, M.P. (the late) 10th Rep., App. VI.
De la Warr, Earl, 3, p. 217; 4, p. 276
De L'Isle, Lord, 3, p. 227
Denbigh, Earl of, 4, p. 254; 6, p. 277; 7, pp. 196-8, 552
De Ros, Lord, 4, p. 317
De Tabley, Lord, 1, p. 46
Devon, Earl of, 3, p. 216; 9, part 2, 403
Devonshire, Duke of (Bolton), 3, p. 36
Devonshire, Duke of (Hardwicke), 3, p. 41
Digby, G. W., 10th Rep.,App. I., p. 520
Digby, R. G. W., 8, p. 213
Dilke, Sir C. W., 2, p. 63
Dillon, Viscount, 2, p. 31
Dod, W., 3, p. 258

Dormer, C. C., 2, p. 82
Dryden, Sir H., 2, p. 63
Effingham, Earl of, 3, p. 223
Egerton-Warburton, R. E., 3, p. 290
Eglinton, Earl of, 10th Rep., App. I., p. 1
Egmont, Earl of, 7, p. 232
Ellacombe, Rev. H. T., 5, p. 323
Exeter, Marquis of, 6, p. 234
Eyston, C. J., 3, p. 260
Farington, Miss, 6, p. 426
Fawkes, A., 7, p. 509
Field, Rev. E., 5, p. 387
Filmer, Sir E., 3, p. 246
Finch, G. H., 7, p. 511 ; 8, p. 640
Fingall, Earl of,10th Rep.,App.v.
Fitzgerald, Sir G., 3, p. 246 ; 5, p. 321
Fitzhardinge, Lord, 4, p. 364
Folkes, Sir W. H. N., 3, p. 247
Fortescue, Earl, 3, p. 220
Fortescue, Hon. G. N., 2, p. 49
Frank, F. B., 6, p. 448
Frere, G. F., 7, p. 518
Gage, Lord, 3, p. 223
Gataere, E. L., 10th Rep.,App.IV.
Gawdy Family, The, 10th Rep., App. II.
Gell, C. P., 9, part 2, p. 384
Gore, J. R. O., 2, p. 84 ; 4, p. 379
Graham, Sir J., 6, p. 319; 7, p. 261
Grey-Egerton, Sir P., 3, p. 244
Griffith, Miss C., 5, p. 405
Grove, S., 5, p. 360
Gunning, Sir H., 3, p. 248
Hailstone, E., 8, p. 636
Hare, Sir T., 3. p. 250
Harvey, John, 1, p. 62
Harvey, J., 2, p. 89
Hatherton, Lord, 5, p. 294
Herries, Lord, 1, p. 45
Hertford, Marquis of, 4, p. 251
Hill, Rev. T. S., 10th Rep., App. IV.
Hopkinson, Rev. F., 3, p. 261

Ingilby, Sir H., 6, p. 352
Isham, Sir C., 3, p. 252
Jersey, Earl of, 8, p. 92
Kilmorey, Earl of, 10th Rep., App. IV.
Knightley, Sir R., 3, p. 254
Lansdowne, Marquis of, 5, p. 215 ; 6, p. 235
Lansdown, Marquis of, 3, p. 125
Lawson, Sir G., 3, p. 255; 5, p. 305
Lawson, Sir H., 4, p. 367
Lechmere, Sir E., 5, p. 299
Leconfield, Lord, 6, p. 287
Lee, J. H., 3, p. 267
Lefroy, T. E., 1, p. 56
Legh, W. J., 3, p. 268
Leicester, Earl of, 9, part 2, p. 357
Leigh, Lord, 2. p. 49
Leighton, Sir B., 2, p. 65
Leighton, Stanley, M.P., 10th Rep., App. IV.
Le Strange, H., 3, p. 271
Lloyd, S. Z., Esq., 10th Rep., App. IV.
Lothian, Marquis of, 1, p. 14
Lowndes, G. L., 7, p. 537
Lucas, Baroness, 2, p. 4
Luttrell, G. F., Esq., 1, p. 56, 10th Rep., App. VI.
Lyttelton, Lord, 2, p. 36
Macaulay, Colonel, 4, p. 397
Macclesfield, Earl of, 1, p. 34
Mackeson, H. B., 2, p. 91
Mainwaring, Sir Philip T., 10th Rep., App. IV.
Majendie, L.,5, p. 321
Malet, Sir A., 7, p. 321 ; 7, p. 428
Manchester, Duke of, 1, p. 12 ; 8, part 2, pp. 1-141
Manning, Cardinal, 5, p. 570
Manning, Rev. C. R., 10th Rep., App. IV.
Manners, Earl, 9, part 2, p. 375
Marlborough, Duke, 8, p. 1

Marsh, T. C., 3, p. 274

Maskelyne, N. Storey, Esq., M.P.

Maxwell, Sir J. S., 10th Rep., App. I., p. 58

Middleton, Earl of, 1, p. 44

Mildmay, Captain St. J., 7, p. 590

Mildmay, Sir H., 5, p. 307

Moray, C. S. H. D., 10th Rep., App. I., p. 81

Molyneux, W. M., 7, p. 596

More, R. Jasper, 10th Rep., App. IV.

Morrison, A., 9, part 2, p. 406

Mostyn, Lord, 1, p. 44 ; 4, p. 347.

Mount Edgecumbe, Earl of, 2, p. 20

Muncaster, Lord, 10th Rep., App. pt. 4

Myddleton-Biddulph, Col., 2, p. 73

Napier, Col., 1, p. 56

Northampton, Marquis of, 3, p. 209

Northumberland, Duke of, 3, p. 45 ; 6, p. 221

Orlebar, R., 3, p. 274

Ormonde, Marquis of, 9, part 2, p. 126, 10th Rep., App. v.

Othen, Miss, 3, p. 277

Parkinson, J. Lechmere, 10th Rep., App. IV.

Peake, F., 2, pp. 92-3 ; 3, p. 277

Pembroke, Earl, 9, part 2, p. 379

Phelips' MSS., 3, p. 281

Phelips, W., 1, p. 57

Pinecoffin, J. R., 4, p. 374 ; 5, p. 573

Plowden, W. F., Esq., 10th Rep., App. IV.

Portsmouth, Earl of, 8, p. 60

Powis, Earl of, 10th Rep., App. IV.

Prescott, Mrs., 2, p. 97

Prideaux. R. W., 5, p. 423

Puleston, Sir R., 2, p. 65

Pyne, 9, part 2, p. 493

Raffles, T. S., 6, p. 468

Ranyard, A. C., 5, p. 434

Ridgway, M., 4, p. 401

Ripon, Marquis of, 5, p. 294 ; 6, p. 243

Rogers, J. J., 2, p. 98 ; 4, p. 405

Rutland, Duke of, 1, p. 10

Rye, W., see Gawdy Papers

Sackvile, Lord, 7, 249

Salisbury, Marquis of. 3, p. 147 ; 3, p. 199 ; 5, p. 261 ; 6, pp. 250-7, 182 ; Calendar, pt. 1, 1883

Salvetti Correspondence, 11th Rep., part 1

Salwey, Alfred, 10th Rep., App. IV.

Sewell, Rev. W. H., 10th Rep., App. IV.

Shaftesbury, Earl of, 3. p. 216

Shirley, E. P., 5, p. 362

Shrewsbury, Earl of, 1, p. 50

Skrine, H. D., 11th Rep., part 1, (Salvetti Correspondence)

Sneyd, Rev. W., 3, 287

Southwark, Bishop of, 3, p. 233

Spencer, Earl, 2, p. 12

Stafford, Lord, 10th Rep., App. pt. 4

Stewart, Capt., of Alltyrodyn, 10th Rep., App. pt. 4

St. Germain's, Earl of. 1. p. 41

Strachey, Sir E., 6, p. 395

Strickland, W. C., 5. p. 329

Sutherland, Duke of, 5, p. 135

Tenterden, Capt., 6, p. 572

Throckmorton, Sir N. W., 3, p. 256, also 10th Rep., App. IV.

Tollemache. John, 1, p. 60

Torrens, W. T. M., 2, p. 99

Towneley, Colonel, 4, p. 406

Townsend, Marquis, 11th Rep., App. part 4

Trelawny, Sir J. S., Bart., 1, p. 50

Turner, W. H., 2, p. 101

Underwood, C. F. Weston, 10th Rep., App. I., p. 199

Verney, Sir A., 7, p. 433
Vidler, J. W. C., 10, p. 14
Walcot, Rev. John, 10th Rep., App. IV
Wallingford, Corporation, 6, p. 572
Webb, Rev. John, 7, p. 681
Westminster, Marquis of, 3, p. 210
Westmoreland, Earl of, 10th Rep., App. IV
Wharncliffe, Lord, 3, p. 224
Whitgreave, F., 1, p. 61

Wilbraham, G. F., 3, p. 292 ; 4, p. 416
Willis, Mrs., 2, p. 103
Wilson, M., 3, p. 293
Wilson, Sir Erasmus, 5, p. 304
Winchelsea, Earl of, 1, p. 14
Winchester, Corporation, 6, p. 595
Winnington, Sir T., Bart., 1, p. 53
Woodford, 9, part 2, p. 493
Wrottesley, Lord, 2, p. 46
Wykeham-Martin, P., 6, p. 465
Wynne, W. W. E., 2, p. 103
Zetland, Earl of, 1, p. 44

APPENDIX VII.

A Short Antiquarian Directory.[1]

ENGLAND.

" Academy," Wellington Street, Strand

Ancient Buildings, Society for the Protection of, 9, Buckingham Street, Strand, W.C. Secretary, Thackeray Turner

Andover Archæological Society

Antiquaries, Society of (Archæologia). Assistant Secretary, W. St. John Hope, Burlington House, W.

Archæological Institute of Great Britain and Ireland. See Royal

Arundel Society, 19, St. James' Street, S.W. Secretary, Douglas H. Gordon

" Athenæum," 4, Took's Court, Chancery Lane, E.C.

Banbury Field Club

Bath Field Club

---

[1] As this is the first time anything of this sort has been attempted my readers must pardon mistakes and omissions. Any correction and additions will be thankfully received.

Batley (Yorks) Antiquarian Society

Bedfordshire Archæological Society

Bedfordshire Notes and Queries, edited by F. A. Blaydes, Shenstone
Lodge, Bedford

Beverley Antiquarian Society

Bradford Antiquarian Society

Bristol and Gloucestershire Archæological Society. Secretary, Rev.
W. Bagelcy, Matson, Gloucestershire; entrance fee, 10s. 6d.;
subscription, 10s. 6d.

British Archæological Association, 32, Sackville Street, W.
Hon. Secs., W. de Grey Birch, F.R.S.L., and E. P. Loftus
Brock, F.S.A.

Buckinghamshire Architectural and Archæological Society, Ayles-
bury. Publishers of *Records of Bucks*

Burton-on-Trent Archæological Society

Cambrian Institute, 1853

Cambridge Antiquarian Society. Hon. Sec., Rev. S. S. Lewis,
Corpus Christi College, Cambridge [in which are merged the
Cambridge Camden Society and the Ecclesiological Society]

Camden Society, 25, Parliament Street, S.W.

Chaucer Society, 67, Victoria Road, Finsbury Park, N. Director,
F. J. Furnivall; Hon. Sec., W. A. Dalziel

Cheshire Notes and Queries. Editor, Rev. R. Blacker, 26, Meridian
Place, Clifton, Bristol; 5s. 5d. yearly; Kent, 23, Paternoster
Row

Chester and North Wales, Archæological and Historic Society for.
Secretary, H. Taylor, Curzon Park, Chester

Chetham Society (Lancashire and Cheshire). Subscription, £1
yearly. Secretary, J. E. Bailey, F.S.A., Stretford, Manchester

Clifton Antiquarian Society. Secretary, W. H. Hudd, F.S.A., 94,
Pembroke Road, Clifton, Bristol; subscription, 10s. 6d. yearly.

Cornwall Royal Institute

Cotteswold Field Club

Cumberland and Westmoreland Antiquarian and Archæological
Society, Kendal

Cumberland and Westmoreland Association for the Advancement of
Literature and Science. Secretary, J. B. Bailey, Eaglesfield
Street, Maryport

Derbyshire Archæological and Naturalists' Society. Hon. Sec.,
Dr. Cox, Barton le Street, Malton, Yorks

Devon, Cornwall, and Somerset Note Book; or, the Western Antiquary;
edited by W. H. K. Wright, Borough Librarian, Plymouth

Devon (North) Diocesan Architectural Society

Devonshire Association

Durham and Northumberland Archæological and Architectural Society. Secretary, Captain E. A. White, Durham; subscription, 5s.

Early English Text Society. Director, F. J. Furnivall; Hon. Sec., W. A. Dalziel, 67, Victoria Road, Finsbury Park, N.

East Anglian; or, Notes and Queries of Suffolk, Cambridge, Essex, and Norfolk. Secretary, Rev. C. K. Evelyn White, Christ Church, Chesham; Publishers, Messrs. Pawsey & Hayes, Ipswich

English Dialect Society. Secretary, W. W. Skeat, 2 Salisbury Villas, Cambridge

English Historical Review, edited by the Rev. M. Creighton, M.A.; Longmans, Green, & Co.; 5s. per quarter

Exeter Architectural Society. Secretary, Rev. J. L. Fulford, Woodbury, Exeter

Folk Lore Society. Publishers, Satchell & Co., 12, Tavistock Street, W.C.

Genealogical and Historical Society

Genealogist, The, edited by W. D. Selby. Bell and Sons, York Street, Covent Garden

Gloucester Notes and Queries, edited by the Rev. W. H. Blacker, 26, Meridian Place, Clifton, Bristol

Harleian Society. Hon. Secs., George J. Armytage, F.S.A., and H. Paul Rylands, Heather Lea, Claughton, Birkenhead; Publishers, Mitchell & Hughes, 140, Wardour Street, W.

Huguenot Society. Secretary, R. S. Faber, 10, Oppidan's Road, N.W.

Index Library. Editor, W. P. W. Phillimore; Publisher, Clark, 4, Lincoln's Inn Fields; 2s. monthly

Index Society. Secretary, H. B. Wheatley, Society of Arts, John Street, Adelphi

Isle of Man Antiquarian Society

Kent Archæological Society. Secretary, ..., Museum, Maidstone

Lancashire and Cheshire Antiquarian Notes

Lancashire and Cheshire Antiquarian Society. See also Chetham Society

Lancashire and Cheshire Record Society. Secretary, J. P. Earwaker, 108, Portland Street, Manchester

Leicestershire Architectural and Archæological Society. Secretary, Col. Bellairs, The Newarke, Leicester; subscription, 10s.

Lincoln and Nottingham Architectural and Archæological Society. Secretary for Lincoln, Rev. J. C. Hudson, M.A., Thornton Vicarage, Horncastle; Secretary for Nottingham, H. Gee, Lincoln Circus, Nottingham

Lincolnshire Archæological Society, A. R., 1844

Lincolnshire Notes and Queries. Editors, E. L. Grange and the Rev. J. C. Hudson, M.A.; 5s. 4d. per annum post free; Publisher, W. K. Morton, High Street, Horncastle

Lincolnshire Topographical Society, 1841

Liverpool Archæological Society

London and Middlesex Archæological Society. Secretary, Tho. Milbourn, 8, Dane's Inn, Strand; entrance fee, £1. 1s.; subscription, £1. 1s.

London Library (lending), St. James' Square; 100,000 vols., including Standard, Legal, and Historical Works. Secretary, R. Harrison; subscription, £3 per annum

London Topographical Society. Secretary, H. B. Wheatley, Society of Arts, John Street, Adelphi

Maidenhead and Thames Valley Antiquarian Society

Manchester Literary Society

Manchester Notes and Queries. Editor, J. H. Nodal, City News Office, Manchester; subscription, 6s. 6d. per annum

Manx Note Book

Middlesex County Record Society. Secretary, J. C. Jeaffreson

Midland Antiquary, published by Cooper, 107, Corporation Street, Birmingham

Miscellanea Genealogica et Heraldica. Editor, Dr. J. J. Howard, LL.D., F.S.A.; Mitchell and Hughes, 140, Wardour Street, W.; subscription, 10s. 6d.; but 7s. 6d. is charged for the title, contents, and index of each vol.

Newbury and Berkshire Field Club

Newcastle-on-Tyne Society of Antiquaries

New Shakspeare Society, University College, Gower Street, W.C. Director, F. J. Furnivall; Secretary, K. Grahame, care of Messrs. Trübners & Co., 57 and 59, Ludgate Hill, E.C.

Norfolk and Norwich Archæological Society. Secs., Rev. W. Hudson, Prince of Wales' Road, Norwich, and Rev. C. R. Manning, Diss Rectory

Norfolk Antiquarian Miscellany (just discontinued, as being no longer necessary, owing to the appointment of the Rev. W. Hudson to the Secretaryship of the Norfolk and Norwich Archæological Society). Editor, W. Rye; Publishers, A. H. Goose & Co., Norwich

Northampton Architectural Society. Secretary, Rev. T. C. Beasley, Dallington, Northampton

Northamptonshire Notes and Queries, edited by . . . ., Dryden Press, Northampton

North Country Lore-Legend, Monthly Chronicle of (illustrated), 6*d.* monthly. Issued by Newcastle *Weekly Chronicle*. [Excellent and very cheap.]

Northern Notes and Queries, edited by the Rev. A. W. C. Hallen, care of David Foulis, Castle Street, Edinburgh

North Riding (Yorks) Record Society. See Yorkshire

Notes and Queries. 4, Took's Court, Chancery Lane, E.C., ; 4*d.* weekly

Nottingham. See Lincoln

Numismatic Society. Sec., B. V. Head, 22, Albemarle Street, W.

Oxford Architectural Society. Secretary, F. A. Dixey, Wadham College, Oxford

Oxford Ashmolean Society. Secretary, E. B. Poulton, Wykeham House, Banbury Road, Oxford

Oxford Historical Society. Secretary, C. F. Madan, Bodleian Library

Oxfordshire Historical and Architectural Society

Palatine Note Book, The

Penzance Antiquarian Society

Photographing Relics of Old London, Society for, 112, Albany Street, N.W. Hon. Sec., A. Marks

Pipe Roll Society. Hon. Sec., J. Greenstreet, 16, Glenwood Road, Catford, S.E. Subscription, £1. 1*s.* yearly.

Royal Archæological Institute of Great Britain and Ireland, 17, Oxford Mansion, Oxford Street, W. Secretary, H. Gosselin

Royal Historical Society, 11, Chandos Street, W. Secretary, P. Edward Dove, Esq., F.R.A.S., 28, Old Buildings, W.C.

Scarborough Archæological Society

Selden Society. Secretary, P. E. Dove, 23, Old Buildings, Lincoln's Inn

Shropshire Archæological Society

Société de l'Histoire de Normandie. Secretary, M. Ch. Legay, care of Ch. Métérie, 11, Rue Jeanne d'Arc, Rouen (British Museum Press Mark, " Ac. 6890 ")

Société des Antiquaires de Normandie, Caen (British Museum Press Mark, " Ac. 5320 ")

Somerset Record Society. Hon. Sec., Rev. J. A. Bennett, South Cadbury Rectory, Bath

Somersetshire Archæological and Architectural Society

St. Alban's Architectural and Archæological Society. Secretary, Rev. Canon Davys, Wheathampstead Rectory, St. Alban's; Local Hon. Sec., Rev. H. Fowler, Lemsford Road, St. Alban's ; annual subscription, 10*s.* 6*d.*

St. Paul's Ecclesiological Society, the Chapter House, St. Paul's Churchyard, E.C. Secretary, W. H. White

Suffolk Institute of Archæology. Secretary, Rev. F. Haslewood, St. Matthew's Rectory, Ipswich

Surrey Archæological Society, 8, Danes Inn, Strand, W.C. Secretary, Thomas Milbourn; entrance fee, 10s.; subscription, 10s.

Surtees Society. Secretary, Canon Raine, York

Sussex Archæological Society

The Reliquary. 2s. 6d. quarterly; Publishers, Bemrose & Sons, 23, Old Bailey, E.C., and Derby; Editor, Rev. J. C. Cox, LL.D., F.S.A.

Walford's *Antiquarian Magazine* (just discontinued). Sets of 12 vols., £3. 3s. nett; Publishers, Redway, York Street, Covent Garden, W.C.

Warwickshire Historical and Archæological Society

Western Antiquary. See Devon

William Salt (Stafford) Archæological Society

Wiltshire Archæological and Natural History Society

Wiltshire Topographical Society

Worcestershire Diocesan Archæological Society, A. R.

Yorkshire Antiquarian Club

Yorkshire Archæological and Topographical Society

Yorkshire (North Riding) Record Society. Editor, Rev. Dr. Atkinson, Danby Parsonage, Yorks

Yorkshire Notes and Queries, edited by J. Horsfall Turner, Idel, Bradford

SCOTLAND.

Ayr and Wigton Archæological Society

Gaelic Society of Inverness

Glasgow Philosophical Society

New Spalding Club, 1887. Secretary, P. J. Anderson, Aberdeen

Scottish Historical Society, 1886. Secretary, Thomas G. Law, Signet Library, Edinburgh

Scottish Notes and Queries, printed by W. Jolly & Sons, 23, Bridge Street, Aberdeen; published by D. Wyllie & Son, Aberdeen; No. 1 issued June, 1887; monthly

Scottish Text Society, 1883. Secretary, Rev. Walter Gregor, M.A., LL.D.; Treasurer, William Blackwood, Esq., 45, George Street, Edinburgh

Index.

₊ In this Index are incorporated many direct Handbook references to subjects treated on in the works on the Public Records by Cooper, Sims, Thomas, Ewald, and Phillimore, with which I have not dealt with in the foregoing book.

The works named in italics with * prefixed are the short titles of books issued by the Rolls Series of Chronicles and Histories, of which details are printed on pp. 150-7.

Abbeys, 17

Abbreviations used in pedigrees, 11

    ,,   Works on [Sims. xiii.] and see Chassant's "Dictionaire des Abbrevations"; also Duffus Hardy's list in "Registrum Palatinum Dunelmense." vol. iv.

Abbrevatio Placitorum, 45 [Cooper, i. 232, 396, for analysis of, showing small proportion printed]

Accountant, General (Chancery), 50 [Sims, 445]

Accounts, Public, Observations made by Commissioners on, in 1691, Camb. Univ. MSS., D.d. xiii. 13, 16, and 33. For list of books contained in the Audit Office, see 1800 Rep. p. 133 b

Acknowledgment of Royal Supremacy, 17, 68

    ,,   Office [Sims, 445]

Acknowledgments by Married Women, 1834, &c., Index in Long-room at R. O. See p. 115

Acts of Parliament, 58. See Parliament

    ,,   Account of the Statutes [Coles, i. 124-206]

Administrations, 91; Form of, 141

Admiral, The office of an, 62 n.

* Admiralty, Black Book of (Twiss)

Admiralty. See Maritime Courts Log Book

    ,,   Instance Court of, report on Records of, 1800 Rep. p. 304

    .,   Records, 62, 63 n.; see also Navy and Cinque Ports [Sims, 82, 440-1]

Ad quod damnum, Inquisition, 32, 95 [Sims, 131]

    ,,   See Inquisitions

Advowson, Descent of, 15

African Co., Royal, List of books of. See 7th Rep. App. II. p. 58

Agarde. See "Repertorie of Records"

Agarde's Index to early Rolls, Digest. See 24th Rep.; and in Appendix to 41st Rep pp. 2-4

Agenda Books, continuation of Memorandum Rolls of Exchequer [Sims, 108]

Agents, List of Record, 108

Agincourt, List of those at, 60

Aid and Subsidy Rolls, 29. See Subsidy [Sims, 45]

Aids, Book of (1327), 31

Alien Priories, 67, 68 [Sims, 82] Accounts of, 21 Edw. I.—22 Edw. IV.— **Record Index, Nos. 16 and 17**

„ Subsidies, 84

Alienation Office [Sims, 445] Accounts of, Jas. I.—**Record Index, No. 18**.

Aliens, 84. See Merchant Strangers ; also " List of Foreign Protestants in England, 1618-88 " [Cooper], and recent works by Moens

Almain Rolls, from 18 Edw. I. See 2nd Rep. p. 45. Inventory in **Record Index, No. 19**

" Alumni Etonenses," 8

Ambassadors. See Treaty Rolls, Treaties, Births, &c., Registers kept by [Sims, 373]

„ Accounts of. There is a descriptive catalogue from 23 Edw. I.—Edward IV. ; and of Warrants, from Henry VIII.—31 Eliz., at the Record Office. Also see French Ambassador, Diplomatic Documents

America, Genealogy of, 4 [Sims, 82, 309-315 332, 373 ; Phillimore, 188]

„ Licenses to go beyond seas, 20th Rep. p. 131

American War, Claims of Loyalists, 15th Rep. p. 3

„ Documents relating to, 24th Rep. p. 85

" Ancient Charters, &c., Calendars of " (Ayloffe), 33

Ancient Demesne Rolls [Sims, 82-85]

„ Miscellanea of Exchequer, 2 Ric. I. See Queen's Remembrancer, 291

Anglo Saxon Charters, 62, 63

„ Chronicle, 13

* *Anglo Saxon Chronicle (Thorpe)*

* *Annales Monastici (Luard)*, 67 n.

Annales, Rotuli. Pipe Rolls, q. v.

Annual Register, Index, 4

Antiquaries, Society of, 24 n.

Apothecaries, 433

Apparel, Documents relating to, temp. Eliz.—**Record Index, No. 20.**

Appearance Books in Actions. See Ewald, p. 58

Archbishops, Lives of (Hook), 61

„ Registers, 126 [Sims, 425]

Archdeaconry Registers, Reports on. See 1800 Rep.

Arches Court of Canterbury, Registry of. See 1837 Rep. p. 263 [Sims, 69] ; and see **Record Index, No. 411**

Archives of the Simancas, 59

Armour. See Meyrick's " Illustrations of ancient Arms and Armour," Boutell and Viollet le Duc

Arms. See Herald's College [Sims, 294, &c.]

Army, 62 ; and see War Office

„ See Foreign Mercenaries

Army Accounts, 48 Henry III.—Elizabeth.—**Record Index, No. 27**
   ,,   ,,   id.—14 Hen. IV., Calendar.—**Id., No. 28**
  ,, Lists, 61, 62 [Sims, 81, 436, 437]
Array, Commissions of. temp. Hen. III. [Sims, 434]
Ashmolean MSS. (Bodleian), Index to the catalogue of (Macray), Oxford, 1866. [See Sims, 446]
Assize Rolls, 51 [Sims, 54, 55, 71]; Calendar of Dockets to.—**Record Index, No. 34**
Association Rolls [Ewald, 58]
Athenæ Oxon., 7
  ,, Cantab., 7
Attainders, 54-61 [Sims, 140, &c.] See Baga de Secretis
Attorneys, Admissions, &c., from 1656.—**Record Indexes, Nos. 35-46**; and see [Ewald, 59]
  ,, Law Lists, 8
Audit Office, For list of books contained in, see 1800 Rep. p. 133 b
  ,, Collectors of, 24th Rep. p. 67
  ,, List of documents open to public, 28th Rep. p. 111 [Sims, 446-7]
Augmentation Bag, Contents of, Hen. VIII.—Edw. VI.—**Record Index, No. 50**; also printed 9th Rep. p. 244
  ,, Court, descriptive slips, Hen. VIII.—Philip and Mary.—**Record Index, No. 51**
  ,,   ,, Index to Proceedings. Hen. VIII.—Edw. VI., transcript of Additional MSS., 21, 291.—**Id., No. 52**
  ,,   ,, See Palmer's Index, vol. cxxxv.
  ,,   ,, Records of [Sims, 61]
  ,, Office, 69 [Sims, 447]
  ,, Charters from, 34
Awards, Enclosure, 19 n., and see Enclosure Awards
Ayloffe's Calendars (Duchy of Lancaster Records), 15.—**Record Indexes, Nos. 53 and 54.**

* *Bacon's Opus Tertium, Opus Minus, &c. (Brewer)*
Baga de Secretis, 54, 61
Banco, Placita de (Common Plea Rolls). For Indexes see Agarde's Indexes
Bank of England, pig-headed obstinacy of Officials of, 81
Bankruptcy Deeds. Before 1831 at Bankruptcy Commissioner's Office, afterwards on Close Roll (Ewald)
Baronages, 9 n.
Baronetages, 102 [Sims, 178, 186, 190-4, 300]; see also "Chaos" in Foster's Baronetage
Baron's Letter to the Pope (1301), Report on 8th Rep. p. 185
Barristers [Sims, 430; and see Foster's Men at the Bar]
Barrows, British. See "A Record of the Examination of Sepulchral Mounds in various parts of England," by the Rev. Canon Greenwell, Clarendon Press, 1877
Beatson's Political Index, 8-61. Register of M.P's., id.
* *Becket, à, Thos., Materials for a History of (Robertson and Sheppard)*
* ,,   ,, *Life of, in Icelandic (Magnusson)*

Bells, Books on, 21 n.
" Bibliographers' Manual " (Lowndes), 16
" Bibliotheca Britannica " (Watts), 8
Bills and Answers. See Chancery, Exchequer, Star Chamber
Bishops, Letters Patent relating to, Separate Roll from 1727, 2nd Rep. p. 44
   ,,   Appointments of, on Patent Roll, Chas. II., 46th Rep. p. 1
   ,,   Possessions, 70. See Parliamentary Surveys
   ,,   Registers, 71 ; Registries, 3
   ,,   Succession of, Dates, and see Stubbs' " Registrum Ecclesiæ
              Anglicanæ"
   ,,   Temporalities, 70. Index Edw. I. to Commonwealth.—**Record
              Index, No. 71 and 72.**
   ,,   Transcripts, 75 ; (printed) 75 n.
* *Black Book of Admiralty (Twiss)*
Black Book of the Exchequer. See Exchequer.
Blanch Silver, 29
Board of Green Cloth [Sims, 61, 330, 447]
Bodleian Library, 136 [Sims, 447]
Boldon Book, 23 [Sims, 6 ; Cooper, i. p. 226]
Bondsmen in Blood. See Villeins
Book of Aids, 31
Border History. Bell's MS. in Library of Dean and Chapter of Carlisle.
               See 1837 Rep. p. 289 b
       ,,   Correspondence as to, Hen. VIII.—James I., 41 fo. vols.
               1837 Rep. p. 78 b; and see Domestic State Papers
       ,,   Also see " Northern Notes and Queries "
* *Bracton de Legibus, &c. (Twiss)*
Bracton's Note Book (Maitland), 62
Brasses, 20 ; Works on, 20 n.
Brevia Regia. See Ewald, 62
* *Britanie, Le Livere de Reis de, &c. (Glover)*
British Museum, 118
       ,,   Chart of Reading-room of, 116
Bulls. See Papal Bulls
Burgi, Firma (Madox), 19
Burial Registers, 73
Bursar's Account (Monastic), 17

Caerlaverock, Siege of, 60
Calais, Rolls and Survey from Hen. VIII., 20th Rep. pp. 78-9
" Calendarium Genealogicum" (Roberts), 86 ; now being continued by Hart,
     87
Calendarium Rot. Chartarum, 32
Calendar, Roman Church [Sims, 472]
Calendars and Indexes in Record Office, on 31st Decr. 1879, List of, 41st Rep.
     appendix II. ; also see 24th Rep.
* *Cambrensis (Giraldus) Chronicle (Brewer and Dimock)*
Cambridge, Report on Documents as to, in Record Office, 20th Rep. p. 136
       ,,   Cooper's Athenæ Cantab, 7-16
Canadian Archives, Report on, 43rd Rep. p. 91

* *Canterbury (Gervase of) Chronicle (Stubbs)*

* *Capgrave's Chronicle (Hingeston)*

Capite, Tenants in, 86

Cardinals' Bundles. Inquisitions of Monasteries surrendered to Wolsey [Ewald, 62].—**Record Index, No. 75.**

Carew Papers in Lambeth Library (1515-1603.)—**Id., No. 76**

Carlton Ride. Notice of Records at [Sims, 449]

Carnarvon, Record of. See Record of Carnarvon

Cartæ Antiquæ, 33 [Sims, 30 ; Ewald 63 ; Cooper i. 318.]—**Record Index, No. 79**

Carte Papers, 60. Carte's Calendar of the Gascon, Norman, and French Rolls [Cooper i. 305]. Final Report on, 32nd Rep.

,, (Bodleian) Transcripts relating to Ireland.—**Record Index, No. 81**

Carthusians, List of, 8

Cartularies. See Chartularies

Castle-guard Rents, Vol. as to, 31 Chas. II.—3 James II. See 1837 Rep. p. 195 a

Castles, 18, 18 n.

Cathedral Libraries, 70

Cathedrals, 70

Catholics. See Roman Catholics

Cecil Papers, 60

Cells of Foreign Abbeys. See Alien Priories

Cemetery Registers, 82 [Sims, 372]

Certificates, Funeral, 7

,, of Birth, Death, &c., 78, 82

Chamber, Accounts of the Treasurer of the, Hen. VII.—Eliz.—**Record Index, No. 88.**

Chancellors, Lives of the (Campbell), 61

,, Lists of [Sims, 330]

Chancellors' Rolls, 24

Chancery, Court of, Records, 48 [Sims, 62]

,, Enrolment Office, 50

,, Printed Calendar of, (Eliz.) 49 [Cooper, i. 354]

,, MS. Calendar of, 49

,, Bill Books, 49.—**Record Index, Nos. 91-100.**

,, Six Clerk's Records, 49

,, Records in Filaciis, 55

"Chancery Files," 55 n.

,, Miscellanea of, 55

Chandos Peerage Case, 9

Channel Islands, Descriptive Slips, Edw. III.—Hen. VIII.—Eliz.—**Record Index, No. 90.**

Chantries, 16, 69 [Sims, 62]

,, Certificate of, Hen. VII.—**Record Index, Nos. 128 and 129**

Chapter-house Records, Inventory of [Sims, 449.]—**Id., No. 102**

Charitable and Religious Gifts, 72 [Ewald, 64]

Charities, 21 n.

,, Decrees relating to, 43 Eliz.—8 Geo. II. ; Index Locorum.—**Record Index, No. 103.**

Charles I., Calendar of Surveys of Estates of, 102

"Chartæ Privilegia et Immunitates," temp. Hen. II.; 92 pp. fo. under this title were published by Government, some years ago, and it sells for 4s.

"Chartarum Calendarium Rotuli," 32

Charter-house, Scholars at, 8

Charter MS. Calendar, Ric. III.—Hen. VIII. Chancery Index, No. 107; and see Palmer's Indexes, vol. i.

,, Rolls, 33-95 [Sims, 79; Cooper, i. 294]

,, ,, Calendar, 32

,, ,, (Norman), 33 n.

Charters, 32 [Sims, 30]

,, Form of, 144

,, Oxford, 33

,, Lambeth, 125

,, Saxon, 33

,, Royal, 34

,, Ayloffes, 15

,, Enrolled on de Banco Roll, 34

,, ,, on Queen's Bench and Close Roll, 34

,, ,, on Charter Roll, 33

Charters of Duchy of Lancaster, Wm. II.—Edw. III.—Record Index, No. 112

,, ,, Hen. I.—Hen. VI.—Id., No. 113

Chartularies, Family [Sims, 28], 17, 66, 67. At Record Office, see 8th Rep. p. 135 [Sims, 14]

Chase, 18

Chetham's Library [Sims, 450]

Chirographers' Office [Sims, 451]

Chivalry, Court of [Sims, 65; Ewald, 64]

,, Collections as to, Lincoln's Inn MSS. xi. (Hales MSS.)

Christ's Hospital, Scholars at, 8

Chronicles, Early, 13

,, Monastic, 17, 66, 67

* Chronicles of Edw. I. and II. (Stubbs)

* Chronicles of Ric. I. (Stubbs)

* Chronicles of Steph., Hen. II., and Ric. I. (Howlett)

* Chronicon Angliæ, 1328-1388 (Thompson)

Chronicon Petroburgense (Cam. Soc.)

Chronological Register (Beatson), 8

"Chronology of History," by Sir H. Nicolas, 8vo., 1835

Church Calendar [Sims, 472]

,, Dignitaries [Sims. 418]

,, Goods Certificates and Inventories, 16, 16 n., 71; List of, 7th Rep. p. 315, and 9th Rep. p. 237, and Record Index, No. 117

,, Lands, Calendar of Bargains as to, temp. Commonwealth, Palmer's Index, vol. 80

,, Lands. Descriptive Slips, Edw. II.—Jas. I.—Record Index, No. 118

,, Registers, 74-79

Churchwardens' Books. 12 [Sims, 388]. See also Overall's "Accounts of the Churchwardens of St. Michael, Cornhill. 1456-1608," 1869

Cinque Ports, Report on Records of the, 1800 Rep. p. 258 a; 1837 Rep. p. 223

Cinque Ports, Vol. as to the, [? Register Book of the Clerk of the Jurators of Romney] in Library of Catharine College, Cambridge; 1837 Rep. p. 338 a

* *Cirencester (Richard de) Chronicle (Mayor)*

Circuits, 51. Reports of the various Clerks of Assize as to their Records, 1800 Rep. p. 237 a, 211, &c.

Circuits, see Midland, &c.

City Records, 19

Civil Lists, 63

,, Servants, 63

,, War Papers, 58, 59

Clare, Honour of, Household Accounts, Descriptive Slips, Edw. II., III.— **Record Index, No. 120**

Clarendon State Papers, 60

* *Clergy, Peacock's Repressor of overmuch blaming of the (Babington)*

Clergy, Records as to, 15, 16, 64 [Sims, 416, 422]. See Ministers

Clerical Subsidies, 15, 64, 83

Clerks of the Peace, 19 ; Enrolment with [Sims, 35]

,, ,, Reports of the Records of the, 1800 Rep. p. 261; 1837 Rep. p. 225

Close Rolls, 97 [Sims, 33, &c. ; Cooper i. 300, 415]

,, John to 1828, Descriptive Indexes to 84 vols.—**Record Index, No. 125**

,, See Palmer's Indexes, vol. ii.

,, Transcripts of, 1 Edw. I.—31 Hen. VI., in 27 vols. See **Record Index, No. 126**

,, Abstract of Deeds enrolled on, temp. Hen. VIII., with Index Locorum ; Palmer's Indexes, vol. lxxi.; ditto, 22-25 Chas. II., id. vol. lxxii.

Coasts. Remarks (1549) as to the Bearings and Soundings of the Coasts of England, France, Norway, and the Gulf of Venice, Cambridge University Library, D.d. iii. 40

Codex, Kemble's, 13

* *Coggeshall (Ralph de) Chronicle (Stevenson)*

Coins. See Mints

"Collectanea Genealogica" (Foster), 7

College of Arms. See Heralds' College

Colleges, Charities, Hospitals, &c. Calendar of Certificates, Hen.VIII.—Edw. VI.—**Record Index, No. 128**

,, Chantries, Hospitals, &c., Particulars for Sale of.—**Id., No. 129**

,, Registers of [Sims, 390, 422]

Colonial Documents, List of, from 1813, 22nd Rep. p. 48. See Canada

,, Office, List of Calendars of, 24th Rep. p. 73

Colonies. See Plantations and East India Company

Commission, Court of High. See Courts

,, Exchequer Depositions by, 47

,, of Grace (Ireland), Calendar of Grants under, 1684-8.—**Record Index, No. 130**

Common Pleas or Common Bench, 46 [Sims, 34, 67]

Commons, House of [Sims, 157]

,, Doctors. See Doctors' Commons

Commons (Common Lands), Enclosure of, 19

Companies, Trading, 70

Compositions for First-Fruits (from 1538).—**Record Index, No. 132**

    ,,   Tithes, 72

Compotus of a Manor, 95 : a Monastic, 17

Concealed Lands, Hen. VIII.—Jas. I.—**Record Index, No. 133**

  ,,   ,,   Entries of Returns of Commissioners from 3-17 Eliz. See 1837 Rep. p. 158 b

Concord Book, 39

Confirmation Rolls [Sims, 134 ; Cooper i. 310]

    ,,   ,,   Inventory, 1 Ric. III.—1 Chas. I. See 4th Rep. p. 134

    ,,   ,,   Calendar, 1 Ric. III.—1 Chas. I.—**Record Index, No. 135**

    ,,   ,,   List of Abstracts and Indexes of, 1880 Rep. p. 92

    ,,   ,,   Index to. Ric. III.—12 Jas. I. See 1837 Rep. p. 111

Consistory Courts, Reports on Records of. See 1837 Rep.

Constables' Roll and Court [Sims, 45, 65]

Contractions. See Abbreviations

Controlment Books, 53

    ,,   Rolls. Minutes of the chief Proceedings in Crown Causes, from 1 Edw. III.

Conventual Leases, 69

Copyholds, 93 [Sims, 85]

Coram Rege (or Crown Plea) Rolls, 44

Cornwall, Duchy of [Sims, p. 445]

    ,,   ,,   Records. See 30th, 31st, 32nd, and 33rd Rep.

    ,,   ,,   Inquisitions of, Post Mortem [Cooper, i. 386]

Coronation Rolls. Jas. III. to present time (except Chas. I. and Geo. III.) See 1837 Rep. p. 111 (now about to be published by Mr. Selby ?)

· ,,   Rolls and Accounts [Sims, 82, 87 ; Ewald, 66]

Coroners' Rolls. 19 ; City, 53, 54 [Sims, 55, 95]

Corporations, Documents of [Sims, 385, &c.,] and see Index to Reports on Historical MSS. Commission at p. 160, 163

Cotemporary Sovereigns, &c., of Europe, Table of, Nicolas' "Chronology of History," 346

* *Cotton (Bartholomew de) Chronicle (Luard)*

Cottonian MSS. Account of the Formation, Contents, and Catalogues of App. F 5 to 1st General Rep. p. 67 [Cole, i. 28-43]

Coucher Books, 67

Councils, Chronological Lists of, Nicolas' " Chronology of History," 201

    ,,   Alphabetical, id., 254

County Bags. Various Documents as to Counties. See 21st Rep. Appendix

    ,,   ,,   Defined. 1819 Rep. p. 195

    ,,   Histories. For Lists of, see Anderson's Book of British Topography, London, 1881

    ,,   (Palatine), Records [Sims, 401]; Reports on Records of, 1800 Rep. p. 253; and 1837 id. p. 219

    ,,   Placita, 46

    ,,   Records and Societies, 19. The Middlesex County Record Society are publishing those for the county; J. C. Jeaffreson. Esq., is editor. 2 vols. are published. The Yorkshire Archæological and Topographical Associations are also doing good work. The Surtees Society has long looked after Durham and York; the Chetham after Lancashire and Cheshire

County Records, Return of Registrars on, 1837 Rep. p. 281

Court of Commission on Forfeited Estates, 61

" Court of High Commission, Notices of the," by J. S. Burn; 8vo. J. Russell Smith, 1865

Court of Requests. See 8th Rep. p. 167, and 24 Rep. p. 53 [Sims, 74]

,,   Star Chamber [Sims, 75]

,,   Survey [Sims, 76]

,,   the Verge [Sims, 73]

,,   Wards and Liveries [Sims, 76]

,,   Wards, 88

,,   Rolls (Manor), 5, 19, 93-5 ; Leet Rolls, 59. Collection of those at Record Office.—**Record Index, No. 1412**

Courts of Chivalry. See Chivalry

Covenant Book (Fines), 39

* *Coventry (Walter de), Historical Collections of (Stubbs)*

Crenellate, Licenses to, 18

Crime. See "History of Crime in England," 2 vols., 1873-6, by L. O. Pike ; also see Circuits, Coram Rege Roll, Felons' Goods, &c.

Criminal Matters, 44 n.

CRIMINAL PROCEEDINGS, &c., 51

Criminals' Papers and Petitions, 53

Crown Causes. See Controlment Rolls

,,   Lands, &c., 101 [Sims, 6, 60-62, 80-87, 319.] Temp. Commonwealth Index, see Palmer's Indexes, 78 and 79 ; also a fo. Rep., issued 1792

,,   Leases, 101, 101 n. See Leases

,,   Plea Rolls, 31, 52, 85. See Coram Rege Rolls

Crusade Roll [Sims, 13]

Curia Regis Rotuli, 44

Cursitor's Office, 50 [Sims, 454]

Customs, 84 ; Accounts of Collectors of (Edw. I.—Jas. I.)—**Record Index, No. 144**

" Custom Revenue in England, History of the," by H. Hall, 8vo. 1885

Custos Brevium Office [Sims, 454]

,,   Sigilli, Accounts of the, Edw. VI.—Eliz.—**Record Index, No. 145.**

Cyphers, diplomatic, Hen. VIII.—Geo. II., 3 fo. vols. of. 1837 Report, p. 78 a

Dates, Short Glossary of [Sims, 498-503]. See also "Medii Ævi Kalendarium" (Kalendar, from 10th to 15th century), R. T. Hampson, 2 vols. 1841 ; " L'art de Vérifier les Dates ;" " Handybook of Rules and Tables for verifying Dates" (J. J. Bond), 1866 ; " Chronology of History" (H. Nicolas), 1835 ; "Jubilee Date Book" (Selby), 1887

Deaths. See Parish Registers

De Banco Rolls, 8, 9

,,   ,,   Charters enrolled on, 34

Decrees and Orders. See Chancery, Exchequer, &c.

Deeds in Treasury of Receipt of Exchequer, Hen. I.—Chas. II.—**Record Index, No. 177**

,,   ,,   ,,   (Diocesan Deeds). **Id.**, No. 178

,, enrolled. See **Record Index, Nos. 183-4-5.**

,, enrolled in Common Pleas, 34

,, Index 15. See Charters

,, of Surrender, 17, 68

,, of Wards and Liveries. See 6th Rep. pp. 1-87

,, Various, Hen. I.—Jas. I.—**Id.**, No. 179

Deer Parks, English (Shirley), 18 n.

Delinquents' Estates (Commonwealth). Palmer's Indexes, vols. lxxiv. lxxviii. lxxix.

Demesne. See Ancient Demesne

,, Roll (Annual) [Sims, 82, 85]

Denization, Letters of, 96

Denmark and Sweden, Report on Royal Archives of, Deposition by Commission. See 45th, 46th, and 47th Reps.

Depositions by Commission (Exchequer), Collections of, 1 Eliz.—4 Jas. II. printed 38th, 39th, and 40th Reps. Indexes to.—**Record Index, Nos. 190-3**

DESCENT OF LAND, &c., 85

Dialogue of the Exchequer, 30

* _Diceto (Ralph de), Chronicle (Stubbs)_

Dictum de Kenilworth [Sims, 71 n.]

"Diem clausit extremum," Form of Writ of, 139

Digby MSS. Transcripts (1605-1695.)—**Record Index, No. 196**

Diplomatic Cyphers, Hen. VIII.—Geo. II., 3 fo. vols. of, 1837 Rep. p. 78 a

Disentailing. Before 1834, by recoveries; after, on Close Rolls, 44

Dispensation Rolls from 37 Eliz. [Ewald, 68]

Dispensations. Marriage [Sims, 32, 361-4]

Dissenters, 73; Registers [Sims, 370, 384]

Dissolution Records, 68

Dissolved Monasteries' Deeds, 34

Dividends, Unclaimed, 3 n.

Docket Rolls, 24. See Doggett

Docks, 30 n.

"Doctors' Commons, its Courts and Registries," by G. J. Foster, 8vo. 1869

Doctor Williams' Library, 79

Documents, Catalogue of, in Exchequer, 1 Hen. I.—38 Hen. VIII.—**Record Index, No. 197**

,,   ,,   Formerly on Treasury of Receipt (Exchequer), Calendar, 45th Rep. p. 283

,, MSS. See Treaties

"Documents illustrative of English History, &c." [Coles, 29-56 n.]

DOCUMENTS RELATING TO THE SUBINFEUDATION, SALE, AND TRANSFER OF LANDS, 21

Dodsworth MSS. (Bodleian, Oxford), Index to first 7 vols of, 25 n.

Doggett or Docket Books (indexes to Common Law Judgment Rolls), 24, 46; before Exchequer, 1 Eliz. 2 b, 1656, and C. P. 1692

,, Book, Great, 53

,, Rolls, before Q. B., 1692.

Domesday, 14-21 [Sims, 1, 9]

Domesday. See Boldon Book, 23

Domesdays (so called) of St. Paul's, Norwich, Ipswich, Chester, Winton, &c., 23

Domestic State Papers, 7 ; and see Appendix for list of printed, p. 157

"Dominibus et Puellis, Rotuli de " (Grimaldi), 4to., 1830, 85

Domus Conversorum (Rolls House). See Jews. Accounts of the Keeper of, Edw. III.—Eliz.—**Record Index, Nos. 201-2**

Douay Diary, 73 n.

Dowager Queens. See Queens Dowager

Duchy of Cornwall. See Cornwall

„ of Lancaster. See Lancaster

Dugdale's Monasticon, 7

    ,,   MSS. in Ashmolean Library [Sims, 446]

    ,,   Origines, 61

    ,,   Peerage, additions to. " Coll. Top. et Gen.," i. p. 51

* *Durham (Symeon de), Chronicle (Arnold)*

Durham Fines, 18

    ,,   Bishop's Registry Records, Report on, 16th Rep. p. 44 ; various Indexes and Calendars of, 30th Rep. p. 44 ; 31st Rep. pp. 42, 112; 32nd Rep. pp. 264, 301; 33rd Rep. p. 43 ; 34th Rep. p. 163 ; 35th Rep. p. 76 ; 36th Rep. p. 1 ; 37th Rep. p. 1 ; 40th Rep. p. 480 ; 40th Rep. p. 480 ; 44th Rep. p. 310 ; 45th Rep. p. 153

    ,,   Palatine Records [Sims, 401]

Dutch Churches in London and Norwich, Registers, 76

* *Eadmer's Historia Novorum (Rule)*

Earl Marshal, Collections as to, Cambridge University MSS. M.m. vol. xii.

    „   Marshal's Court [Sims, 44, 65, 73, 82]

Easter Day, How to find, from A.D. 1000 to A.D. 2000. See Nicolas' " Chronology of History," 58-78

East India Company [Sims, 372, 436-8]

    ,,   ,,   Calendar of Papers relating to, 1588-1755, 1819 Rep. p. 363

    ,,   ,,   Deeds as to, 1702 ; 1837 Rep. p. 158 c

ECCLESIASTICAL AND MONASTIC RECORDS, 64 [Sims, 422]

Ecclesiastical Commission, Account of Receipts, Eliz.—8 Jas. I.—**Record Index, No. 206**

    ,,   Courts. See [Sims, 68, 343]

    ,,   ,,   Report on Records of the, 1800 Rep. p. 304, and 1837 Rep. p. 257

Ecclesiastica, Taxatio, 15

Ecclesiasticus, Valor, 15

* *Edward the Confessor, Lives of (Luard)*

Edward VI. Grammar Schools, Particulars of the Estates of, 1837 Rep. p. 208 b

Election Cases in Parliament, 1625-8. Maynard's MSS. (Lincoln's Inn) lxxii.

Elections. See Pawns, Parliamentary Poll Books ; and [Cooper, i. 323]

Ely, Domesday of, 23

Emblems of Saints. See Husenbeth on, 1880

Emigrants, Lists of, 4

Encaustic Tiles, 20 n.

Enclosure Awards, 19 n. [Sims, 402], Index from 1756-1853.—**Record Index, No. 296**

Enclosure of Common Lands, 19
Engraving and Illuminating, Methods of [Phillimore, 61]
Enrolment Office (Chancery), 50 [Sims, 458]
Entry Books of the Exchequer, 84
Equitium Regis, Accounts of, 17 Edw. I.—20 Jas. I.—**Record Index, No. 207**
    ,,   ,,   Calendar of Documents relating to, Edw. I.—Jas. I. ; 20th
            Rep. p. 118
Equity. See Chancery
Equity side of the Exchequer. See Exchequer, 47
Error, Writs, &c., in [Ewald, 70]
Escheators' Accounts, 87, Hen. III.—Jas. I.—**Record Index, No. 211**
    ,,   ,,   See Inquisitions post mortem
Escheat Rolls [Sims, 61, 96, 123]
Escheats, 40, Rolls [Sims, 41, 42, 60, 82, 87, 105, 125]
Essoin Rolls [Sims, 68 ; Ewald, 70]
Estreats, 24 [Ewald, 70]
Etonensis, Alumni, 8
* *Evesham, Chronicle of (Macray)*
Exannual Rolls [Sims, 71]
Excerpta è Rotulis Finium, 14
Exchequer Chamber (Chancery), 47
    ,,   of Pleas Court (Common Law), 46
    ,,   ,,   Doggett Books of, 48
    ,,   ,,   Index to Orders and Decrees, 47
    ,,   Depositions by Commission, 47
    ,,   Dialogue concerning the, 30
    ,,   Madox's History of the, 24
    ,,   Martin's Index to the, 96
    ,,   Norman (Stapleton), 24
    ,,   of Receipt (Treasury) Records, 84.   See 24th Rep., and Thomas'
            "Ancient Exchequer of England," 97
    ,,   ,,   ,,   Miscellanea, 29
    ,,   ,,   ,,   Originalia, 96
    ,,   ,,   ,,   Remembrancer of the, 48
    ,,   ,,   ,,   Issue or Liberate Rolls, 47
    ,,   ,,   ,,   Black and Red Books of the, 30
    ,,   Queen's Remembrancer of the, 29
Exon Domesday, 23 [Sims, 5 ; Coles, i. 208]
Extents, from Chas. I.   See 1837 Rep. p. 118 a
    ,,   of a Manor, Specimens of.   See 1800 Rep. 145-6
" Extracta Donationum " [Sims, 139]
Eyre Rolls, 51

Fabric Rolls, 67
Fairs and Markets, Grants of, 1 John—22 Edw. IV., Palmer's Indexes, vol.
    xciii. and cvi.
Fairs, Grants of, 18.   See Pat. Roll and Charter Roll
Family Chartularies [Sims, 28]
Fasti Eccl. Anglic. (Le Neve and Hardy), 65
Fee-farm, Grants of, Hen. VIII.—Jas. I., Palmer's Indexes, vol. cxxiii.

Fee-farm Rents, 69 [Sims, 61, 105, 119, 402]
,, ,, Particulars for the sale of, Commonwealth.—**Record Indexes, Nos. 219, 220, and 221**
Fees, Knights'. See Knights' Fees.
Feet of Fines, 35 ; form, 36 [Cooper, ii. 428, i. 428 ; Sims, 132]
,, Concords of, 39
,, Covenant Book, 39
,, Entries, 38
,, King's Silver Books, 39
,, Published and MS. Calendars, 40-1-2
Felons' Goods, Accounts of, &c., Hen. IV.—Hen. VI.—**Record Index, No. 222.** See Crime
Fen Registers. See [Sims, p. 36]
Fens, Various papers as to the Great Level of the, 1800 Rep. p. 174
,, Lands set out for the King's use, temp. Chas. I., id. p. 191 a
Feodaries, Certificates of Inquisitions, 87 n.
Feodorum, Liber, 30
Festivals, Calendar of Saints' and other [Sims, 504-11] ; and see Nicolas' "Chronology of History," 116-177
Feudal Service (Inquisitions, &c.), Henry III.—Hen. VIII. **Record Index, No. 223.**
Filacer's Office [Sims, 456]
Filaciis, Chancery Records in, 55
Fine Rolls (Licenses to alienate, &c.), 31-32 [Sims, 98 ; Cooper i. 308]
,, Index Locorum, 7 to 18 Hen. III.—**Record Index, No. 227**
,, Calendar, 1 Edw. I.—7 Edw. II.—**Id., No. 228**
,, ,, Edw. V.—Chas. I., Palmer's Indexes, vols. lxxv.-lxxvii.
Fines and Recoveries. See Feet of Fines
,, Feet of. See Feet of Fines
Finibus, Rotuli de oblatis et, 31, 32
Firma Burgi. See Fee Farm
First-Fruits, 65 [Sims, 417]
,, and Tenths Offices, Report on Records of, 1800 Rep. p. 209 a
,, ,, Accounts of, Hen. VIII.—Jas. I.—**Record Index, No.256**
FISCAL RECORDS, SUBSIDY ROLLS, &c., 82
Fisheries, 63 n. [Sims, 60, 82, 105]
,, Collections as to, from Rolls of Parliament, Close, and Patent Rolls, Cambridge University MSS. D.d. xi. 71
,, Treatise on fishing in the north seas temp. Jas. I. ; Maynard MSS. (Lincoln's Inn) lxvii
,, Petitions as to, from 1542-1761. See 1819 Rep. p. 363
,, Reports on the Herring Fisheries, 1819 Rep. p. 113
Fleet Marriages, 80 [Sims, 376]
* *Flores Historiarum. See Wendover*
Fœdera (Rymer's), 55
,, Account of [Cooper, ii. 89, 144]
Foreign Estreats, Records of the Clerk of. See 1837 Rep. p. 201
,, Mercenaries, Account of Disbursements for, in wars against France and Scotland, 1800 Rep. p. 172 a

Foreign Merchants, Edw. I.—Hen. VIII., 20th Rep. p. 126
,,   ,,   Documents relating to, Edw. I.—Hen. VIII.—**Record Index, No. 259**
,,   Monasteries. See Alien Priories
,,   Office, Calendars of; 24th Rep. p. 73
,,   Protestants [Sims, 367]
,,   State Papers, 59
" Foreshore, History of the " (Stuart Moore), now publishing
Forest Proceedings. Descriptive Inventory, John—Chas. I. Printed 5th Rep. pp. 46-59
,.   Rolls, 102
Forests, Royal, 18, 18 n. [Sims, 82, 100, 104, 109, 319, 403, 422]
,,   ,,   See Crown Lands
,,   Placita [Sims, 57]
Forfeited Estates, 58-61 [Sims, 32, 82, 108, 111, 112, 140, 319]
,,   ,,   Accounts of. Descriptive Slips, Edw. 2.—Ric. II.—**Record Index, No. 264**
,,   ,,   Accounts of. Descriptive Inventory, Geo. I. Printed, 5th Rep. App. II.—**Id., No. 265**
Forfeitures [Sims, 140]
Forms of Legal Documents. See App. p. 139; also Madox's " Formulare Anglicanum," and " Archæological Journal," 1865, p. 58
Fortify, License to, 18
Fortress, History of a, 18. See Castle
Founder's Kin. [Sims, 179, 392]
Franciscana Monumenta, 155
Fraternity, Guild or, 71
Freemen's Roll, 4
Free Warren, 18 [Sims, 60, 79, 105]
French and English History, 36th Rep.
,,   Ambassadors in England, from 1509-1714, 37th Rep. p. 180
,,   Archives, Report on Documents in, relating to England
,,   Despatches of (same date), 39th Rep. p. 573
,,   ,,   Transcripts, 43rd Rep. p. 609
,,   Rolls, Hen. V., Calendar of, 44th Rep. p. 543 (see Norman Rolls) [Cooper. i. 305; Sims, 99]
,,   Wars. See French Rolls
* *French Chronicles of Great Britain* (*Hardy*)
Friars. See Monastic Establishments
* *Friars Minors, " Monumenta Franciscana"* (*Brewer and Howlett*)
Funeral Certificates 7, 82 [Sims, 279, 284]
Furniture, Church, 71

Gaol Delivery Rolls, 51
Garter, Order of, the Statutes of, Cambridge University MSS., D.d. x. 47, 59, and xi. 47; and see 1837 Rep. p. 13 a
Gascon Rolls, Calendared by Carte [Ewald, 74; Cooper, i. 305; Sims, 100]
Gauger's Accounts. Edw. VI.—Eliz.—**Record Index, No. 275**
General Register Office, 80
,,   Surveyors, Court of, 101

Gentleman's Magazine, 4-5, 54

Gentry, Lists of [Sims, 326]

Gilds, 16-70

Glossaries, Mediæval. See Ducange; also "Promptorium Parvulorum," (Way, Cam. Soc.)

,, Excellent ditto in H. T. Riley's "Memorials of London and London Life" (Norman), 42nd Rep. pp. 453-472

,, Pipe Roll, vol. iii. of Pipe Roll Society

Grace, Commission of (Ireland), Calendar of Grants under, 1684-8.—**Record Index, No. 130**

Grammar Schools. See Edw. VI.

Grants, Enrolments of, 32

GRANTS FROM THE CROWN PRIVILEGES, TITLES, &c., 95

Grants, Particulars for (dissolved monasteries), 7, 32

,, See Particulars for Grants, 68-69

,, passing Great Seal, 10 Edw. V.—Ric. III.—**Record Index, No. 278**

Great Roll of the Pipe. See Pipe Roll, 24-29

,, Rolls, or Pipe Rolls, 24-29

,, Seal. See Custos Sigilli

Green Cloth. See Board of Green Cloth

,, Wax, Surveyor of, Office [Sims, 466]

Gretna Green Marriages, 80

* *Grosseteste, Bishop, Letters of, (Luard)*

Guild Certificates, 70

,, ,, List of, Ric. II., Palmer's Indexes, vol. cxi.

Guilds, 16-70 [Sims, 62, 385]

Hall (Hubert), "History of the Customs," 84

Hanaper Office [Sims, 458]

Hardwick MSS. (Brit. Mus.) Calendar of, 4to., 1794

Hargrave MSS. (Brit. Mus.), Calendar of (Ellis), 4to., 1818

Harleian MSS., Account of the formation, contents, and catalogues of, App. F 6 to General Rep. p. 71 [Coles, i. 44-115]

Hatfield, MSS. at, "Cecil Papers," 60

Hawking, &c. Rotulus Expensarum Falconum Austurcorum et Venatorum temp. Regis Edwardi, 6 and 7 Edw. I., 1837 Rep. p. 186 a

,, Roll of Expenses of the Kings', 12 Edw. I., id. p. 18 b

Hearth Tax Rolls, 84

* *Henries, the illustrious, Capgrave's Chronicle of (Hingeston)*

● *Henry III., Historical Letters (Shirley)*

* *Henry V., Memorials of (Cole)*

* *Henry VI., Memorials of, Reign of (Williams)*

* ,, *Letters, &c., illustrative of French Wars, temp. (Stevenson)*

* *Henry VII., Materials for History of (Campbell)*

* ,, *Memorials of (Gairdner)*

Heraldic Collections [Sims, 159]

Heralds' College, 129. Report on Records of [1800 Rep. pp. 82-3; 1837 Rep. p. 106; Sims, 452, &c.]

,, Visitations, 6-132; printed, 132

,, College, Dublin [Sims, 452]

Herring Fisheries. See Fisheries

* *Higden (Ralph) Polychronicon (Babington & Lumby)*

High Commission, Court of, Notices of the, by J. S. Burn, 8vo. 1865

* *Historical Letters, temp. Hen. III. (Shirley)*

Historical MSS. Commission, 59. List of Reports of, App. p. 160

" Historical Notes," 1509-1714 (H. S. Thomas), 61

" History from Marble " (Dingley), 7

Homage Rolls (temp. Edw. I.) See Ragman Roll

Home Circuit, Report on Records of the, 1800 Rep. p. 237 a

   ,,   Office, Calendars of, 24th Rep. p. 72

Horses, Rolls as to, temp. Edw. I., 2nd Rep. p. 59. See Equitium Regis

   ,,   Rolls as to the King's, 5-9 Edw. I., 1837 Rep. p. 185 b

Hospital, Bill of Work done at Greenwich, 1698-1709, in City of London
    Library, 1837 Rep. p. 426

Hospitallers, Knight, 67

Hours, Canonical, Nicolas' "Chronology of History," 183-4

House Duty Accounts, Notes on, 1800 Rep. p. 169

Household Rolls, Bishops', 69

   ,,   ,,   Expenses, 69

   ,,   ,,   For reference to Royal, see 1800 Rep. pp. 169, 174 a ;
             and 1837 Rep. p. 14 a

   ,,   ,,   Transcripts, 44 Hen. III.—12 Edw. I.—**Record Index,**
             No. 291

   ,,   ,,   and Accounts, 32 ; Collector of, 20th Rep. p.128 [Ewald,
             75]

   ,,   ,,   and Wardrobe Accounts, John—Geo. III., 14 vols. MS.—
             Record Index, Nos. 289 and 290

   ,,   Officers of the [Sims, 329]

House of Commons, 57-98

   ,,   Lords, 57

* *Hoveden (Roger de) Chronicle (Stubbs)*

How to Compile a Pedigree, 1

How to write the History of a Parish or other Place, 12

Huguenots, A peaceable Survey of the Doctrine of the, proving they are not
    Heretiques, &c., by a Roman Catholic, 4to. MS., York Cathedral ; 1837
    Rep. p. 287 a, (see Huguenot Society in Directory)

Hundred Courts, Rolls of, Descriptive Slips, Hen. III.—Hen. VII.—**Record
    Index, No. 292**

   ,,   Rolls, 31, 100 [Sims, 104 ; Cooper, i. 267, 282]

* *Huntingdon (Henry of) Chronicle (Arnold)*

Hustings Court (London), 91. Rolls of Pleas, 1800 Rep. pp. 175 b, 176 a

Imports. For list of articles imported at Southampton, see " Archæological
    Journal," 1859, p. 343 ; at Lynn (R. Howlett), "Norfolk Antiquarian
    Miscellany," iii. p. 603

Inclosure Awards, 19 n. [Sims 402,] 1756-1853.—**Record Index, No. 296**

Indentures of War, 62 n.

Independents. See Dissenters, 73

" Index Monasticus " (Taylor), 66

Index Society, 4

Indexes, Calendars, &c., at P.R.O. on 31st December, 1879, List of, 41st
    Rep. Appendix II.
  ,,   Table showing relative frequency of surnames under letters
    [Phillimore, 19]
Index Library (promoted by Mr. W. P. W. Phillimore, published by Mr. C. J.
    Clark, 4, Lincoln's Inn Fields; annual subscription £1. 1s.; is
    printing indexes from the Record Office.)
India, List of Commissioners to manage. Board of Control from 1790; 31st
    Rep. p. 367
Indictments. Indexes called "Pye Books," 53
  ,,   Preserved from 1673. As to Indexes, see [Ewald, 76]
Indulgence, Form of a Papal, "Archæological Journal," 1860, p. 250; 1865,
    p. 62 [Sims, 466]
Inner Temple. A Catalogue of printed books and MSS. in the Library, 8vo.
    1833
Innocent's (Pope) Taxation, 64 n.
Inns of Court, Registers of, 8 n. [Sims, 430]
Inquisitio Eliensis, 23 [Sims, 5; Coles, i. 222]
Inquisitiones ad Quod Damnum 32-95 [Cooper, i. 295-6; Sims, 131]
  ,,   Nonarum, 15
  ,,   Post mortem (Chancery and Exchequer), 86; Form, 139 [Sims,
    32, 123; Cooper, i. 332]
  ,,   ,,   Wards and Liveries, 87-8; Various Calendars,
    &c., 87 n.
  ,,   ,,   (Duchy of Lancaster), 41
  ,,   Quo Warranto, 31
Inrolment Office, 50 [Sims, 458]
Inscriptions, Monumental, 12
Instance Court of the Admiralty, 62 n. Report on Records of, 1800 Rep. p. 304
Institution Books, 15, 71 [1556-1836] [Ewald, p. 77]
Institutions, Indexes to, 71. **Record Index, No. 70**
Insurrections in North of England; 1807 Rep. pp. 13 b, 14
Inventions, see Patents, Specifications of, 98, 98 n.
Inventories (Testators'), 93
  ,,   of Church Goods, 16, 71
Ireland, see Carte Papers, and see Index to [Sims, q. v.]
  ,,   Chronicles as to, App. p. 154
  ,,   Registry of Births, &c., 81
Irish Records [Sims, 410]
Iron Mill Forges in St. Leonard's Forest, Sussex, 1800 Rep. p. 185 b
  ,,   Mines in Rutland, Stafford, and Salop, 1689, 1800 Rep. p. 195 a
  ,,   ,,   Bundle containing 7 books of Accounts, temp. Eliz., id. 200 a
Issue Rolls of Exchequer, 84

Jacobites, 54
Jesus, Society of, 73 n.
Jewels. See Royal Jewels
  ,,   of Hen. VI., 1837 Rep. p. 13 n.
Jews, Rolls relating to, 2nd Rep. p. 54
  ,,   Documents relating to the Domus Conversorum, 20th Rep. p. 118

Jews, Descriptive Catalogue of Documents as to, Hen. II.—20 Ed. I., and as
    to the Domus Conversorum, 35 Hen. III.—Ric. III.
  ,, Account of the Keeper of, Edw. III.—Eliz.—**Record Index, No. 201-2**
  ,, Jews' Rolls, 10 Hen. III.—23 Ed. I., 24th Rep. p. 45
  ,, Litterœ obligationœ pro solutionibus debitis Judœis, id.
  ,, Other Rolls as to, id. pp. 182-184
  ,, Roll as to Jews of Lincoln, among Miscellanies of King's
    Remembrancer, 1837 Rep. p. 181 b
  ,, Various References [Sims, 427; Ewald, 77]
  ,, Various Rolls as to (1-7 Edw. I.), 1837 Rep. p. 186 b
  ,, Tallage accounts, &c.; Descriptive Slips, Hen. II.—Edw. I.—**Record
    Index, No. 330.**
  ,, Calendar, **Id.**
Jones' "Index to the Records," 96
Journals of Parliament, 57
Judges (Foss's), 61
  ,, Origines Juridicales, 61
  ,, Chron. Juridicales, 61
Judgments. How signed and entered, 2nd Rep. p. 53
  ,, Indexes and entries to [Ewald, 78]
Jury Lists (Queen's Bench), from 1747
  ,, List (Special); Index, 1794-1842.—**Record Index, No. 339.**
Justices of the Peace [Sims, 332]

Kemble's "Codex Diplomaticus," 13
King's Bench. See Queen's Bench
  ,, Books. Liber Regis. See Valor Ecclesiasticus and [Sims, 418]
  ,, Remembrancer. See Queen's Remembrancer, 29
  ,, Silver Book (fines), 39. From Eliz., give more information than
    bare Indexes
Kirby's Quest, 35 Edw. I., 31
Knights Hospitallers, 67
  ,, Templars, 67, and see work on, by Addison, 8vo. 1852
  ,, Lists of, Le Neve, &c., 6; also [Sims in Index, q. v.], and Metcalfe's
    Knights, Henry VI.—Charles II.
  ,, Fees, 31 [Sims, 82, 97]

Lambeth Library [Sims, 459]
  ,,   ,, Charters at, 33
  ,,   ,, Wills at, 92
Lancashire and Cheshire Records, 87, 92
  ,, Inquisitions Post Mortem, Ric. II.—Eliz., 39th Rep. pp. 533-549
  ,, Records (County Palatine), Inventory 35th Rep. pp. 42-75 [Sims,
    402]
Lancaster, Duchy of, 50; Inventory and List of Documents transferred to
    Record Office, 1860, 30th Rep. pp. 1-43
  ,, Calendar of Royal Charters, 31st Rep. p. 1, and 35th Rep. p. 1;
    36th Rep. p. 161
  ,,   ,, Chancery Rolls, 32nd Rep. p. 331; 33rd Rep. p. 2;
    37th Rep. p. 172

Lancaster, Calendar of, Patent Rolls (4 Ric. II.—21 Hen. VII.), 40th Rep.
    p. 521
        ,,   ,,   Court Rolls, 43rd Rep. p. 210
        ,,   ,,   Privy Seals, 43rd Rep. p. 363
        ,,   Inventory of Ministers' and Receivers' Accounts, 45th Rep. p. 1;
                Index at p. 123
        ,,   Charters of, printed by Hardy, 50
Land Office, now so called (formerly Copyhold, Enclosure, and Tithe Com-
        mission)
    ,,   Revenue, Account of enrolments of, 1800 Rep. p. 172 b
    ,,   ,,   Records of the, id. p. 202
    ,,   Revenue Office [Sims, 460], 1800 Rep. p. 169
    ,,   Tax, Duplicates of the, 1837 Rep. p. 157
    ,,   ,,   Notes on Accounts, 1800 Rep. p. 169
* *Langtoft (Pierre de) Chronicle (Wright)*
Lansdowne MSS., Account and Catalogues of [Coles, i. 116, 123], and
        Appendix F 7 to General Rep. p. 85
Law List, 8
    ,,   Terms, Nicolas' "Chronology of History," 339
        ,,   Table shewing duration of, from Norman Conquest to Wm. IV.
        [most useful], 28th Rep. pp. 114, 139
Lawyers [Sims, 428]
Lay Subsidy Rolls, 15
Leases (Conventual) granted by the Monasteries before the dissolution, 69 ;
                Index Locorum.—**Record Index, 360 and 361**
    ,,   Index, 1677-1822.—**Id., No. 346** ; and see Indexes calendared at p.
                29 of App. to 41st Rep.
Ledger Books, 17 [Sims, 12]
Leet Rolls, 53
* *Leechdoms, Wortcunning, and Starcraft (Cockayne.)   See Medicine*
LEGAL PROCEEDINGS RELATING TO (1) LAND, AND (2) OTHER MATTERS NOT
        CRIMINAL, 44
Le Neve's Catalogue of Knights, 6 ; Fasti, 65
Leet Rolls, 53
Letter Books (City), 19
* *Letters, Historical, temp. Hen. III. (Shirley)*
Letters, Missive. See 1819 Rep. pp. 189, 367 ; and for an abstract, p. 368.
                Stated to be in number about 15,000
    ,,   Inventory, Edw. III.—Chas. I.—**Record Index, No. 366**
    ,,   Royal, 55
* *Letters, Royal and Historical, temp. Hen. IV. (Hingeston)*
Liber Albus (London) 19 ; Decimarum, 72
    ,,   De Antiquis Legibus [Cooper, ii. 326]
    ,,   Feodorum, 30 (Testa de Nevill) [Cooper, i. 258, 266]
    ,,   Niger Scacc., 30 [Sims, 39 ; Cooper, ii. 324]
    ,,   Regis, 65
    ,,   Ruber Scacc , 30 [Sims, 40 ; Cooper, ii. 309, 325]
Liberate Rolls, 84 [Sims, 106 ; Ewald, 80 ; Cooper, i. 308]
Libraries. See British Museum, Bodleian, Cathedral, Lambeth. &c.
    ,,   List of Public. See Appendix to 1st General Rep. p. 176

Licenses, Marriage, 7 [Sims, 363]
„ to pass beyond the sea, Eliz.—Chas. I.—**Record Index, No. 371**
Lincoln's Inn Library [Sims, 460]
* *Lincoln, St. Hugh of, Life of (Dimock)*
Liveries and Wards, Court of, 88
Livings, Presentations to. See Institution Books, 71
Log Books, 63 n. See Navy.
London Gazette. Index to Orders in Council, Proclamations, and other matters published in, 1830-1883. London, 1885, very thick 8vo.
Lord Chamberlain's Office, 101 [Sims, 461]
„ Chamberlain's Records, 101
Lords Lieutenant from Geo. II., 43rd Rep. p. 722
„ House of. 57. See Peers and Peerage Claims.
Lowndes' Bibliographical Manual, 16
Lunacy and Idiocy. Commissions and Inquisitions, from Chas. I. to present time. See 1837 Rep. p. 118 a
„ „ Some Inquisitions de lunatico inquirendo among the Escheat bundles, 1837 Rep. p. 113
„ „ Commissions and Inquisitions of, Chas. I.—1852.— **Record Index, No. 377**

Magna Charta, Facsimiles of, 1819 Rep. vols. iii., v., and vii.
* *Malmesbury's (Wm. of) Chronicle (Hamilton)*
MANORIAL RECORDS, COURT ROLLS, &c., 93
Manors, Routine of Management of, 93 [Sims, 7, 61, 85, 105, 109, 319, 422]
„ Court Rolls, 93. Notes on printed, 94
„ Extents of, 94 : Specimens. See 1800 Rep. p. 145-6
„ Lete Rolls of, 94
„ Compotus of a, 95
Mansion House, 18
Manuscript Commission, Royal, Catalogue of Reports of, 160
* *Manuscripts, Hardy's Descriptive Catalogue of, relating to the History of Great Britain*
Manuscripts, National, Descriptive list of facsimiles of. See 26th, 27th, 28th, 30th, and 31st Reps.
Markets, 18
Maritime Courts, Reports on Records of, 1800 Rep. p. 303
Marriage Licenses, 7 [Sims, 100, 103, 361, 363] See Dispensations
Marriages, 80. See Parish Registers; Dissenters' Registers
Marshal, Earl, Court Rolls, &c. [Sims, 44, 65, 73, 82]
Marshalsea Court [Ewald, 88]
„ Documents as to, 20th Rep. p. 131
„ Proceedings in [Placita Aulæ] 12 Edw. I.—32 Edw. III. 1800 Rep. p. 38 b
„ and Palace Courts. Reports on Records of, 1800 Rep. p. 215 a
„ of the King's Household, &c., Accounts of, 25 Edw. I.—36 Eliz.—**Record Index, No. 379**
Martin's Index to the Exchequer, 96.—**Id., No. 382**
Masters in Chancery [Sims, 461]
Materials for a History of Great Britain [Cooper, ii. 145, 177]

Materials for History of Great Britain, List of Transcripts for, made from
    2 Eliz. [Cooper, ii. pp. 365, 370]

Matriculation Books [Sims, 390-3]

Maynard's MSS. (Lincoln's Inn), 45 n.

Medical Profession [Sims, 432]

Medicine. MSS. Treatises on. See Codrington Library, All Saints, Oxford,
    1837 Rep., pp. 333-4. See Leechdoms

Members of Parliament, 63 [Sims, 132-4, 157]

    ,,   ,,   Scotch (Foster) "Collect. Gen.," ii.

Memoranda on Heraldry (Le Neve), 6

Memoranda Roll of the Exchequer, 48 [Sims, 35, 69, 107 ; Ewald, 81]

    ,,   ,,   ,,   ,,   Report on, 1837 Rep. p. 177.   See Ex-
        chequer, p. 48

Mercenaries. See Foreign Mercenaries

Merchant Strangers, Orders to be observed by, Hales MSS. (Lincoln's Inn),
    lxxviii. No. 81

    ,,   Taylors, Scholars at, 8

Messengers, Royal, Fees, Edw. II.—Hen. VIII.—**Record Index, No. 119**

Methodists. See Dissenters, 73

Metropolitan Cemeteries, 82

Middle Hill MSS. (See Phillipps). " Her. et Gen." viii. 349

Middlesex County Record Society. Mr. J. C. Jeaffreson has issued 2 vols. of
    Records for this Society, and the work is in progress

    ,,   Registry Office, 35

Middle Temple Library [Sims, 467]

Midland Circuit, Report on Records of, 1800 Rep. p. 238 b

Military Offices. See Army, &c.

    ,,   Defences, 62 n.

    ,,   Papers, 62 n.

    ,,   Service. See Knights' Fees, Parliamentary Writs, &c.

Mines, Accounts relating to, Edw. I.—Chas. I.—**Record Index, No. 393**

    ,,   List of Documents as to, from Edw. I., 20th Rep. p. 131

    ,,   Vol. of Collections as to, Hales MSS. (Lincoln's Inn) cx. See Coal,
        Iron

Ministers' Accounts, 18 ; of Lands, &c., in the hands of the Crown, Hen.
        III.—Jas. I., arranged under the places.—**Record
        Index, No. 394**

    ,,   ,,   (Bigod's lands) Hen. III.—Edw. II.—**Id., No. 395**

    ,,   ,,   (Dissolution), 68

    ,,   ,,   Rough Calendar of, 1st Rep. p. 127 : 2nd Rep. p. 42

    ,,   ,,   List of, 20th Rep. p. 84 [Sims, 318, 402 ; Ewald, 81-2]

    ,,   of State, Lists of [Sims, 332]

    ,,   (Plundered) Accounts, 59 [Clergymen]

Mints, Vol. of Collections as to, Hales MSS., Lincoln's Inn, cx.

    ,,   Assays, Indentures, &c., of, Chapter House Records, 1819 Rep. pp.
        194-5-6

    ,,   Indenture between the Queen and Sir Isaac Newton, 1702, as to
        Coining, 1837 Rep. p. 158

    ,,   MS. as to Coins, &c., by Nicholas Tyery, Edw. VI. (?), Cambridge
        University MS., F.f. 11, 22

Mints, Lists of various Accounts, &c., 1584, &c., 1800 Rep. p. 176 a
,,　Roll of Purchasers of Silver for, 54-5 Hen. III., 1819 Rep., p. 187
,,　Book of Charges of the Old Mint-house, 18 Hen. VIII., id. p. 14
,,　Accounts, Henry III.—Geo. II.—**Record Index, No. 402.**
,,　List of Documents from Henry III., 20th Rep. p. 131 [Sims, 380]
Misæ Roll [Sims, 107; Ewald, 82]
Miscellanea of Chancery, 55
,,　　Exchequer, 29
Missals, 67-73
Monasteries, Dissolution or Suppression Records, 68. See Monastic Establishments
* *Monastic Chartularies and Records, p.* 17
Monastic Establishments, 65 [Sims, 10]
,,　Registers, 67
,,　Suppression. See Monasteries, Dissolution, 68
,,　Smaller Documents of the, 1837 Rep. i. p. 12
Monasticon, Anglicanum (Le Neve) 7; Dugdale's, 6, 65; local, 66 [Dodsworth's]
,,　Eboracense, 66
Monasticus, Taylor's Index, 66
Monumenta Franciscana. See Friars Minors, 155
,,　Historica Britannica, 14
Monumental Inscriptions, 82
Monuments, How to date [Sims (quoting Cutts), 282-4]
Mortuary Rolls, 67
Municipal Offices, Index to, 19
Musters, Rolls, 61 [Sims, 103, 434-9]
,,　See Army

Naturalization, Certificates of, before 1844. See Patent Rolls as from 1844 to August, 1870. Indentures. Since August, 1870, the Certificates are registered at the Home Office.
Navy, 62. See Admiralty; Loss; Ships [Sims, 81, 439, 440]; Log Books
,,　&c., Accounts of the, Edw. III.—Eliz.—**Record Index, No. 421**
Necrologies, 67 [Sims, 12]
Nevill, Testa de, 30
Newcourt's Repertorium, 15
Newgate Calendar, 54
New River, Book of Rents of, 1621-2, 1800 Rep. p. 191 b
,,　,,　Account of Rents of, 1627-8, id. p. 191 a
Nicholas, Pope, Taxation of, 64 [Sims, 47; Cooper, i. pp. 283-5]
Nomina Villarum [Sims, 318]
Nonæ Rolls, 15 [Cooper, i. 286]
Nonarum Inquisitiones, 15 [Sims, 46]
Nonconformists, 73
Nonjurors, 72 n.
Non-parochial Registers, 79 [Sims, 365]
Norfolk Circuit, Report on Records of, 1800 Rep.
Norman Rolls (Calendared by Carte) for 1200-5 and 1417, printed by Record Commissioners, 8, 1835

Normandy, Charters, &c., granted in England to various Religious Houses in, 3 vols. MS. entitled " Cartulaire de la Basse Normandie."—**Record Index, No. 430**
* *Normandy, Narratives of the expulsion of the English from*, 1449-50 (*Stevenson*)
Norman Rolls (Hardy). A Calendar of Continuation in 41st Rep. p. 671; and 42nd Rep. pp. 313, 452 [Sims, 102 ; Cooper, i. 305]
  ,,   Glossary of Obsolete French Words on the, 42nd Rep. pp. 453-472
* *Northern Registers, Historical Papers and Letters from* (*Raine*)
Norwich Taxation, 1253, 64 n.
Nova Taxatio, 1318, 64 [Cooper, i. 283]
Numerical Characters, Nicolas' "Chronology of History," 185
Nuncii (Accounts, &c., of Ambassadors), 23 Edw. I.—13 Eliz.—**Record Index, Nos. 441 and 432**
Nuncupative Wills, 90, 91
Nunneries. See Monastic Establishments

Oaths of Allegiance, Papers, Oaths, &c., Rolls of, 1837 Rep. p. 119 a
  ,, of Office, Forms of, 1800 Rep. p. 216 ; Report on ditto, id. p. 99 a
Obits, 67 ; Mawson's, 6
Obituary, Musgrave's, 7
  ,,   Notices, 5
Oblata Rolls, 31, 32 ; 2nd Rep. [Sims, 110 ; Ewald, 15]
  ,,   Transcripts, 1 John—52 Hen. III.—**Record Index, No. 435**
Oblatis et Finibus, Rotuli de, 32
Officers, List of Public [Sims, 329]
  ,, of the Household [Id.]
Offices, Appointments to, Index.—**Record Index, No. 436**
  ,, Grants of, 100 [Palmer's Index, vols. xxxvi., lvii., lxiii., cviii., cix., cx., cxxxvi.
Ordnance, 62 n. See Army
  ,, Accounts relating to, Edw. II.—Jas. I.—**Record Index, No. 440**
Originalia, Jones' [Chapman's] Index to the, 96 [said by Cooper, i. 343, to be very imperfect]
  ,, Rolls (Exchequer), 84-96 [Sims, 111 ; Ewald, 85 ; Cooper, i. 342] 1837 Rep.
* *Oxenede (John de) Chronicle (Ellis)*
Oxford. See Universities [Sims, s. v.]
  ,, Charters, 33
  ,, Foster's Matriculations, 7-16
  ,,   ,,   Alumni, 7-16
  ,, Historical Society, 7
  ,, References to Documents at the Record Office as to, 20th Rep. p. 136
  ,, Register of University, 7
  ,, Wood's Athenæ, 7-16

Palace Courts, [Sims, 73, 461 ; Ewald, 86]. Report on Records of, 1800 Rep. p. 215 a
Palaces. See Royal Palaces
Palatine Counties' Records [Sims, 401]

Palmer's Indexes (mostly Patents from Hen. VII.) Description of. See 20th Rep. p. 184

Papal Briefs (1513-1527) Calendar of, 2nd Rep. p. 190; 3rd Rep. p. 187; 4th Rep. p. 212; 6th Rep. p. 246

,, Bulls, Calendar of, temp. Edw. II., 1837 Rep. p. 12

,, ,, List of original, at Record Office, 5th Rep. p. 45

,, ,, [Ewald, 86]

Papists, 72 n. See [Sims, s. v.] See Roman Catholics

Pardon Rolls, 96, 96 n.

Pardons [Sims, 140]

* *Paris (Matthew) Chronicle (Luard)*

PARISH, HOW TO WRITE THE HISTORY OF A, 12

PARISH REGISTERS, CEMETERY BOOKS, &c., 73

Parish or Town Books, 4, 12

,, Clerk's Society, 3 n.

,, Registers, 35, 72-73; Abstract of 1830 [Sims, 351]

,, ,, Fees on Searchings, 77

,, ,, List of printed, 76, 77 [Sims, 358]

,, ,, Transcripts of, 75

,, ,, Works on, 74 n.

Parks. See Royal Parks

,, English Deer (Shirley), 18 n.

Parliamentary Pawns. See Pawns

,, Petitions; Chancery; Calendar, Edw. I.—Hen. VII., 2 vols. MS.—**Record Index, No. 458**

,, ,, Transcripts, 66 vols. MS.—**Record Index, No. 459, and see Nos. 460 and 461.**

,, ,, Records, the Publication of [Cooper, ii. 1-18]

,, Surveys, 58

,, Writs, 98 (Palgrave's, 102); Account of the Volumes entitled [Cooper, ii. 33-88]

Parliament, Journals of, 57 [Sims, 157]; Votes of, id.

,, Acts of, 57

,, Petitions, 56 [Cooper, i. 325]

,, Rolls of, 56 [Cooper, i. 322]

,, Ryley's Placita Parliamentaria, 56 n., and see [Cooper, i. 319]

,, Summons to, 98

,, Votes and Debates, 57

,, Writs of Election [Cooper, i. 323]

Parochial, non-, Registers, 79

Particulars, 32

,, for Grants, 68-69: Inventory, Hen. VIII.; 9th Rep. pp. 148-309. **Record Index, No. 466**

,, ,, Index Locorum, ditto, 4 vols. MS. referring to the above.—**Id., No. 467**

,, ,, Calendar, Edw. VI., 2 vols., MS.—**Id., No. 468**

,, ,, Index Locorum, Edw. VI.—**Id., No. 469**

,, ,, Calendar, Phil. and Mary—Jas. I.—**Id., No. 470**

,, ,, Index Locorum, ditto.—**Id., No. 471**

,, ,, Descriptive Slips, Edw. VI.—Jas. I.—**Id., No. 472**

Particulars for Grants, Index Locorum, Edw. VI.—**Record Index, No.** 347
    ,,   Leases, 69
    ,,   the Sale of the Estates of Charles I., &c.—**Id., No.** 474
Patent Rolls, 98 [Cooper, i. 297 ; Sims, 114]
    ,, Books (Auditors' and Pell Series), 101
    ,, Palmer's Index to, 100
    ,, Signet Index, 100
    ,, Transcripts, 1-16 Hen. III.—**Record Index, No.** 486
    ,,   ,,   1 Edw. I.—31 Hen. VI.—**Id., No.** 487
    ,, Calendars and Indexes, 99 ; 1-57 Hen. III. and Edw. V.—39 Vict., 59
        vols. MS.—**Id., No.** 488. For detailed references to other Indexes,
        see p. 45 of Appendix to 41st Rep.
Patents, Specifications of [Inventions] [Sims, 81, 83, 111]
Pawns, Parliamentary, from 21 Hen. VIII. See 1837 Rep. p. 118 b
    ,, Bundles relating to Parliamentary Elections [Ewald, 87]
Peace, Justices of [Sims, 83]
Peculiars, Justices of [Sims, 347]
PEDIGREE, HOW TO COMPILE A, 1 ; a specimen, 10
    ,, Abbreviations used in, 11
    ,, Marshall's Index to printed Pedigrees, 6
    ,, Specimens of different modes of setting out [Phillimore, 31]
Peerage Claims, 102 [Sims, 242-279]
Peerages, Grants of, 102
Peer of the Realm, Report on Dignity of, 63
Peers, Creations of, from Rich. III., 47th Rep. p. 78
Pell Accounts. Descriptive Slips, Jas. I.—**Record Index, No.** 496
    ,, Office (see Exchequer and Devon's Issues) [Sims, 461]
    ,, Records, 84
    ,, Rolls [Sims, 450]
Perambulation Rolls of Forests, 18 n., 102 [Ewald, 87]
* *Peterborough (Benedict of) Chronicle (Stubbs)*
Petitions to Parliament, 56 [Cooper, i. 324-331
Petty Bag Office, 50. See 2nd Rep. p. 43 [Sims, 462]
Peveril Court, Records of the, 1661-1849, 16th Report, pp. 43, 44
    ,,   ,,   ,,   ,, Report on 16th Rep. p. 41 [Ewald, 87]
Phillipps', Sir Thomas, MSS. and Press at Middle Hill. See "Her. et
    Gen." viii. p. 349. The MSS. and a set of his prints are now at
    Thirlstane House, Cheltenham, and I believe can be consulted on pay-
    ment of fee of £1. 1s. per diem
Physicians, Roll of College of, 8 [Sims, 433]
Pipe, Records of the Clerk of the, 1837 Rep. p. 198
    ,, Rolls, 24-29 [Sims, 119 ; Ewald, 88 ; Cooper, i. 312, 317]
    ,, Roll Society, 25 n.
Pirates, 63 [Sims, 83]
"Placita Anglo-Normanica" (Bigelow), 44 n.
Placita Communia, 46
    ,, Aulæ, Marshalsea Court [Ewald, 88]
    ,, Coram Rege, Curia Regis, 44 [Sims, 53 ; Cooper, i. 232-257]
    ,, de Quo Warranto, 31 [Sims, 7, 59]
    ,, Terræ, 46

Placita Parliamentaria, 56 n. [Sims, 59 ; Cooper, i. 319]

" Placitorum Abbrevatio," 45

Plantations. Petitions as to, 1819 Rep. p. 363

Plate, Works on, 21 n.

" Pleas of the Crown," by F. W. Maitland, 51

Plundered Ministers' Accounts, 59

* *Poems, Anglo-Latin, Satirical, of the Twelfth Century (Wright)*

Political Index (Beatson's), 8, 61

* *Political Poems and Songs, Edw. III.—to Henry VIII. (Wright)*

Poll Books, 4 [Sims, 323]

,, Tax Rolls, 83 [Sims, 83]

Pope Innocent's Taxation, 64 [Cooper, i. 283]

    ,, Nicholas' ,, 64 n. See Taxatio Eccles., 6, 30, and Cooper, i. 283

    ,, Chronological list of, Nicolas' " Chronology of History," 187

    ,, Alphabetical, id., 197

Popes' Bulls. See Papal Bulls

" Popular Genealogists," 9

Population Abstracts, 74

Ports. See Cinque Ports

Post Mortem Inquisitions, Chancery Exchequer, 85; Form, 139 [Sims, 32, &c., Cooper, i. 332-341]

Præstita Roll, 12 John [Sims, 167 ; Ewald, 89]

Prebendal Church. See Lee's " History of the Prebendal Church of Blessed Virgin Mary of Thame "

Prerogative Court of Canterbury [Sims, 462]

Presentations. See Institutions, 15-71

Prices. See Thorold Rogers' " History of Prices "

Priories. See Monastic Establishments, 65

    ,, Alien, 67, 68 [Sims, 82]; Accounts of, 21 Edw. I.—23 Edw. IV. —**Record Index, Nos. 16 and 17**

Prisoners, Royal, Expenses of, Edw. III.—Hen. VII.—Id., No. **585**

    ,, in Tower, List of, 30th Rep. p. 313

Private Acts, 57 n.

Privileges and Immunities, temp. Hen. II. See " Chartæ Privilegia," &c.

Privy Council, Records of, 101

    ,,   ,, Nicolas, 56

    ,,   ,, Office, 56

    ,,   ,, Registers, 56 n.

    ,, Councillors [Sims, 331]

    ,, Purse, Expenses of, Hen. VIII., by Sir H. Nicolas, 1827

    ,,   ,, Princess Mary, by T. M., 1831

    ,,   ,, Indexes to, &c. [Phillimore, 81]

    ,, Seal. See Signet Office, 100

    ,, ,, Books, 101 ; 12 Eliz.-1834, printed, 7th Rep. App. II. pp. 31, 32

    ,, Seals [Sims, 136 ; Office, id. 462 ; Ewald, 89.] See also as to, p. 51 of App. II. to 41st Rep.

    ,, Signet Office, 100

Probate Act, Form of, 141

    ,, Registries, 5

" Procedure, History of " (Bigelow), 51

Proofs of Age, 88, 89

Protestant Nonconformists, 73

Prothonotaries Office [Sims, 453]

Provisions, Accounts of Expenses for, Edw. III.—**Record Index, No. 599**

Prynne's Records, 45 n.

Public Offices. See " Notes for Materials for the History of Public Depart-
    ments," by F. S. Thomas, fo. 1846, sells for £1. 1s.

 ,, Officers, Lists of [Sims, 329]

 ,, Schools, Lists of Scholars at, 8 [Sims, 398]

Puritans, 73

Pye Books, 53 (Indexes to indictments), 26 [Ewald, 90]

Quakers' Registers, 79 [Sims, 371]

Queen's Bench, 44, 45 [Sims, 34. 70]

 ,, Dowager, Rolls and Records as to Possessions of various, viz.,
    Catherine Parr, Anne of Denmark, Henrietta Maria, Catherine
    of Braganza, 1800 Rep. p. 177 a

 ,, Remembrancer of the Exchequer, 29 [Sims, 107, 459, 463 ; Cooper,
    ii. 331]

 ,, Silver Books (Fines), 18 ; from Eliz., give more information than
    bare indexes

Quo Warranto Rolls, 31, 100 [Sims, 7, 59, 71, 105 ; Cooper, i. 267, 282]

Ragman Roll (or Homage Roll), temp. Edw. I. [Sims, 407, 408 n.]

Rebellions of '15 and '45, 61. See Forfeited Estates, State Prisoners, &c.
    [Sims, 71, 140, 147]

Record and Writ Clerks' Office (Chancery) [Sims, 465]

 ,, of Carnarvon, 2 fo., 1838, a Welsh Record described by [Sims,
    414-5]

 ,, Office described, 8

Records in the Tower, Inventory of, 2nd Rep. p. 2

 ,, (City), London, Nottingham, and Oxford, 19

 ,, in Filaciis, Chancery, 55

 ,, of the Privy Council, 28

Recoveries, 43 [Sims, 34, 132]

Recovery Index, 34, 44 [Ewald, 9]

 ,, Rolls, or Placita Terræ, from 25 Eliz.

Rectors, 15

Rectory, Certificate of Value and Deductions of, Lyson's Environs, i. p. 63

Recusants, 72. See papers. Lists of [Sims, 32, 143-4, 423]

 ,, Rolls, 58, 5th Rep.

Red Book of the Exchequer, 30

Redisseisin Rolls [Ewald, 92 ; Cooper, i. 308 ; Sims, 122]

Refugees [Sims, 145-6, 367, 373.] See also publications of Huguenot
    Society

Regimental Registers [Sims, 388]

 ,, Muster Rolls [id., 435]

Register Bills, 3

Registers, Parish, 73 : Transcripts, 77
    ,,   ,,   printed, 76
    .,   Regimental [Sims, 388]
    ,,   Various, for detailed references. See [Sims, 522-3]
Registrar General's Office, 80
Registrum Roffense, 66 ; Index to Monumental Inscriptions in Thorpe's, printed by Mr. F. A. Crisp, 1885
Registry, Middlesex, 35
    ,,   York, 35
Regnal years [Sims, 484 ; Nicolas' "Chronology of History," 324]
Remembrance Book [Ewald, 92]
    ,.   Roll, 44
Remembrancer of the Exchequer, 29
Repertorie of Records, by Powell, first published in 1622, is said to have been made up out of Agarde's Collections, 1819 Rep. p. 477
"Repertorium Eccles : London," Newcourt's, 15
Report Office, Chancery, 50 [Sims, 445]
"Report on the Dignity of a Peer of the Realm," 63
Reports on Public Records, Deputy Keeper's, *passim*
Requests, Court of, 8th Rep. p. 167 [Sims, 74 : Ewald, 93]
    ,,   Records of, 1800 Rep. p. 39 b : 24th Rep. p. 53
Rex Rolls. See "Coram Rege or Crown Rolls"
* *Ric. III. and Hen. VII., Letters, &c., illustrative of (Gairdner)*
Riders to Issue Rolls of Common Pleas, 1837 Rep. p. 135
Rolls Chapel. See Jews
    ,, House or Domus Conversorum. See Jews. [Sims, 464]
    ,, Records, 87 n.
    ., of Parliament, 56
Roman Catholics, 58, 73 n.
    ,,   ,,   Persecutions of, 58, 72 n., 73 n. See Papists, Marriages of, 80
    ,,   ,,   Lists of, 73 n. [Sims, 81, 349, 355, 381]
    ,, and Church Calendars [Sims, 472]
    ,, Rolls [Ewald, 93 ; Cooper, i. 308]
Rome, Report on researches among Archives of Vatican, 41st Rep. p. 815
Rotuli Annales or Pipe Rolls, 24-29 [Cooper, i. 312]
"Rotuli Curiæ Regis," (Palgrave), 45
"Rotuli de Dominabus," &c. (Grimaldi), 40
"Rotuli de Oblatis et Finibus," 14
"Rotuli Lit. Claus." (Hardy), 98
"Rough Coat Books," containing references to Charter, Patent, and Close Rolls, John—Edw. IV., Index Locorum, **Chancery Index, No. 108** ; and see Palmer's Indexes, vols. xlii.-vii.
* *Royal and Historical Letters, temp. Hen. IV. (Hingeston)*
Royal Descents, 9. See works on this subject by Burke and Foster
    ,, Charters, 34
    ,, Forests, 18, 102
    ,, Jewels, Wardrobes, &c., in the Exchequer of Account, 1800 Rep. pp. 170 a, 171 b, 174 b ; and see (Ewald, 77) ; and see Jewels
    ,, Letters, otherwise Chancery Records in Filaciis, 55

Royal Letters, Index, Ric. I.—Edw. III., 1 vol. MS. An alphabetical index of persons and places, referring by number to the Royal Letters, from 1 to 2,330, printed in 4th, 5th, 6th, and 7th Rep.

,,   ,,   Calendar, Ric. I.—Edw. II., 4 vols. MS., referring to the, numbered from 1 to 4,700. The calendar to the first 2,300 is printed in the Reports

,,   Messengers' Fees, Edw. II. to Hen. VIII.—**Record Index, No. 119**

,,   Palaces (see Castles) Repairs done, temp. Hen. VIII., 1837 Rep. p. 18 b (68 fo. vols. as to Hampton Court !)—**Id.**

,,   Parks and Castles, Description of Books of Accounts as to, 1800 Rep. p. 174 b

,,   Supremacy, Acknowledgments of, 17

Royalist Composition Papers, 59. [Sims, 438]

Rugby Register, 8

Rymer's Fœdera, 55

Sailors, 61. See Navy. [Sims, 341, 439]; Wills of, [Sims, 349]

Saints' Days, Calendar of [Sims, 472]

,,   ,,   Alphabetical List of [Id., 504] ; much fuller list of, Nicolas' "Chronology of History," 124-166

,,   Emblems. See Husenbeth on

* *St. Albani, Chronica Monasterii (Riley)*

* *St. Dunstan, Memorials of (Stubbs)*

St. Paul's School, Register of, 8

SALE AND TRANSFER OF LAND INTER VIVOS, 32

Saxon Charters, 33

Schools, Public, Registers of, 8 [Sims, 398, 432]

* *Science, Neckam's De Naturis Rerum (Wright)*

Scotch Birth and Registry Office, 81

,,   Rolls, Ayloffe's Calendar of, 33 [Cooper, i. 306]

,,   Records [Sims, 523]

* *Scotland, Chronicles as to,* 154

,,   Public Records of [Cooper, ii. 179-191 ; Sims, 404]

,,   Parliamentary Records [id. 193-221]

,,   Notes of other Scotch Records [id. 206-307]

Scrope and Grosvenor Roll [Sims, 66]

Scutage Rolls [Sims, 43, 44]

,,   ,,   Transcript, Hen. III.—**Record Index, No. 635**

Sea, Licenses to cross the. See Licenses

Seal, Great. See Custos Sigilli

Seals, &c., Catalogue, Hen. II.—Eliz.—**Record Index, No. 636**

,,   List of, at Record Office, 20th Rep. p. 91

Sea Ports (Eastern), 63

See, Vacant, Search during a, 15

Seizin, Form of Delivery of, 146

Selden's Collections. See Hale MSS. vols. xii. and xiii. (Lincoln's Inn MS.)

"Sepulchral Antiquities" (Blore), 21 n.

Sequestered Estates [Sims, 141]

Sequestrations, Committee for, 58

Sessions Rolls, Transcripts of early (1596-7), Transactions of Yorkshire Archæological and Topographical Society

Sewers, Enrolments of Laws of, Chas. I.—1714.—**Record Index, No 640**

,, Laws and adjudication of, Papers as to. See 1837 Rep. p. 118 a

Sheriffs, List of, from 31 Hen. I.—4 Ed. III., 31st Rep., p. 262

,, Accounts, Hen. III.—Jas. I.—**Record Index, No. 644.** List of, 20th Rep. p. 133

Ships, 63 n. See Navy, Sailors, and Log Books

Signet Books (Indexes to Patent Rolls, q.v.), 100 [Sims, 134], now being printed by W. Phillimore

,, Bills [Sims, 134]

,, Office, 100 [Sims, 465], vol. of precedents probably for the use of the Keeper of the Privy Seal, Cambridge University MSS. D.d. iii. 53

Sign Manuals [Sims, 134]

Simancas, Archives of the, 59

Six Clerk's Office, 50, and see 2nd Rep. p. 40, and 1837 Rep. p. 126 c [Sims, 465]

Society of Antiquaries, 24 n.

Soldiers, 61 [Sims, 433 ; Regimental Registers of, id., 388]. See Army

,, Families, Regimental Registers [Sims, 388]

Sovereigns of Europe, Table of Cotemporary, Nicolas' "Chronology of History," 346

Specification Rolls, 98. See Patents (Inventions)

Specimen Pedigree, 10

Stannary Courts. Reports on Records of, 1800 Rep. p. 259 a ; 1837 Rep. p. 225

,, ,, Collections as to, Maynard MSS. (Lincoln's Inn) lxiii. pt. 2

Staple Rolls, (as to Statute Staples) [Ewald, 95]

,, ,, Inventory, 27 Edw. III.—39 Hen. VI. Printed Report, ii. Appendix ii. p. 36.

Star Chamber, 54, 54 n.; Notices of the Court of, by J. S. Burn, 8vo. 1870 ; Accounts of Fines, &c. ; Descriptive Slips, Hen. VIII.— Jas. I.—**Record Index, No. 659**

,, ,, Office [Sims, 75, 465]

,, ,, Proceedings of the Court of, Hen. VIII.—**Record Index, No. 660**

STATE PAPERS, &c., 55-59 (Domestic and Foreign), 27, 29

State Paper Office [Sims, 465]

,, ,, List of Published. See Letters Missive ; Chancery Records in Fillaciis

State Prisoners. See Tower

,, Trials, 54, 61

Statute Rolls [Sims, 148, 149 ; Ewald, 96]

Statutes, Staple, from Chas. I. See 1837 Rep. p. 118 a

,, 58. See Acts of Parliament

Strangers. See Aliens and Merchant Strangers, 84

Subsidy Rolls, 5, 15, 64, 82, 83 [Sims, 45]

,, ,, Alien, 83

,, ,, Lay, 15, 83

,, ,, (Clerical), 15, 83 ; Descriptive Slips.—**Record Index, No. 684**

Subsidy Rolls (Clerical and Lay) ; Descriptive Inventory, Hen. III.—Eliz.
    ,,   ,,   (Lay); Descriptive Index of, Hen. III.—Wm. and Mary.—
            Record Index, No. 685.
Summonses to Parliament, 98 [Sims, 81, 152, 184-5]
Summons, Writs of Military, 63
Suppression of Monasteries, 17.   See Dissolution
Surnames, List of Common English. See 16th Rep., Registrar General
Surrender. Deeds of Abbeys, &c., temp. Dissolution, 17 [Sims, 35]
Survey, Court of [Sims, 76]
Surveyor of Green Wax Office [Sims, 466]
Surveys, Parliamentary, 58
Sweden. See Denmark

Taxatio Ecclesiastica (of Pope Nicholas), 15, 64 [Ewald, 98 ; Coles, i. 283]
    ,,   Norwicensis (Innocent IV. in 1253) [Cooper, i. 283]
    ,,   Ecclesiastica (Nicholas IV., 1291) [id.]
    ,,   Nova (1318) [id.]
Taxes. See Land Tax, Subsidies, &c.: also " History of Taxation and
    Taxes in England " (Dowell, Longmans)
Templars, Knight, 67
Temple, Inner and Middle Libraries [Sims, 466-7]
Tenants in Capite, 86
Tenths and First Fruits, 65 [Sims, 417, &c.]
Terms. Table showing the duration of the Law Terms from the Conquest to
    Wm. IV. [most useful] 28th Rep. p. 114
Terriers, 17 [Sims, 32]
Testa de Nevill, 30 [Sims, 42-3, 97 ; Ewald, 98 ; Cooper, i. 258, 266]
Testament, 90
Tiles, Encaustic, 20 n.
" Times," Mr. Palmer is publishing a most valuable Index to the
Tithe Commutation Office, 19
    ,,   Compositions for, 72
    ,,   Map, 19
    ,,   Suits enrolled, 72
Topography, The Book of British, by J. P. Anderson, Satchell, 1881, 6 n.
    Gives a list of all printed topographical works in the British Museum
Tournament Rolls [Sims, 13, 83, 203]
Tower of London, Accounts of the Constable of the, Edw. I.—Hen. VI.—
            Record Index, No. 702
    ,,   ,,   Index to Records of Prisoners in, 30th Rep. p. 313
    ,,   ,,   Other Records, id., 353 [Sims, 467]
    ,,   ,,   Prisoners in, 54
Town Books, 4, 12
Towns, 19
Trade, &c., Petitions as to, from 1542-1761 ; 62 vols. with Calendar in 3
    vols. fo., 1819 Rep. p. 363
Transcripts of Parish Registers, 75 ; of Fines, 43 ; of Pipe Rolls, 25
Treasury (see Exchequer), 84
    ,,   Records of, 97
    ,,   Papers, Calendar, 1557-1714, 4 vols., printed.—**Record Index, No. 704**

Treaties, &c. (England and Foreign Countries), 1 vol. MS. See also Agarde's
    Indexes, vol. xliv.—**Record Index, No. 706**
,,   and Treaty Rolls. See [Sims' Index, 525; Ewald, 99]
,,   Various. See 1837 Rep. pp. 12 14
,,   and Diplomatic MSS., Williamson's Collection of, (40 fo. vols.) Hen.
    VIII.—Chas. II. See 1819 Rep. p. 364
,,   Treaty or Foreign Rolls, from Edw. V., continued irregularly to 19
    Chas. II. See 1837 Rep. p. 111 (no indexes)

Ulnager's Accounts, Descriptive Slips, Edw. III.—Jas. I.—**Record Index, No. 708**
,,   ,,   List of, 20th Rep. p. 135
Unclaimed Dividends, 3 n.
Universities of Oxford and Cambridge, Accounts of the Revenues of the
                various Colleges in 37 Hen. VIII.
                (2 fo. vols.) 1837 Rep. p. 209 a
,,   ,,   ,,   ,,   Documents at Record Office as to, 20th
                Rep. p. 136
University Registers [Sims, 390]

Vacancy Wills, 92
Vacant See, Search during, 15
Valor Ecclesiasticus, 15, 17, 65 [Cooper, i. 348 ; Sims, 47, 418, 423; Ewald, 99]
Valor, Verus, 64.
,,   Vetus, 64 n.
Vardon's Index to the Journals of Parliament, 57
Vascon Rolls [Sims, 100], and see Gascon Rolls, &c.
Vatican MSS., Transcripts of, 59
Venetian Archives, MS., 59
Verge, Court of [Sims, 61, 73]
Verus Valor, The, 64
" Vetus Codex ; or, Black Book of the Tower " [Sims, 61; Ewald, 100]
Vetus Valor, 64 n.
Vicars, 15
Villeins, 46
Visitation Books, 17
Visitations, Ecclesiastical [Sims, 32]
,,   Heralds, 6
Votes of Parliament, 57

* *Wales, Chronicles as to*
Wardrobe Rolls and Accounts (see Household Books), various
      ,,   Calendar of, 2nd Rep. p. 199
      ,,   Various Rolls as to, 1 Edw. I., 1837 Rep. p. 190
      ,,   Rolls, 1-11 Edw. I., id. p. 186 a
      ,,   Books, 1-34 Edw. I., id. p. 192
      ,,   Accounts, 12-16 Edw. III., id. p. 12
      ,,   Books, 17 Edw. II. 2nd Rep. p. 63
War, Indentures of, 62 n.
,,   Office, 61. See Army and 24th Rep. p. 81
,,   See Army

Wardrobe Rolls Book, 4-20 Edw. II., 1837 Rep. pp. 192-3
,,   Accounts of the Queen's, in 1608 (?), Cambridge University MSS., D.d. i. 26
,,   Roll as to Ordinances of, 17 Edw. II., 2nd Rep. p. 63; and see Royal Household Books, &c. [Ewald, 100]
,,   and Household Accounts, Enrolments of, 9 Edw. I.—52 Geo. III.—**Record Index. No. 715**
Wards and Liveries, Court of, 88 [Sims, 76, 454; Ewald, 100]
,,   Marriages (Palgrave), 105
Warrant Books, Calendar of. See 1819 Rep. p. 363
Warranto, Quo, 31
Welsh Records [Sims, 413] Report on, 16th Rep. 35; 17th Rep. 38; Catalogue of, 24th Rep. 54
,,   ,,   Various Reports on. See 20th Rep. 23, 25th Rep. 23, 26th Rep. 32, 26th Rep. 16, 27th Rep. 96, 28th Rep. 6, 29th Rep. 49, 30th Rep. 121, 31st Rep. 169, 36th Rep. 1, 37th Rep. pt. 2, 39th Rep. pt. 1. See also Sims' Index, 525; Ewald, 101; Cooper, i. 307; Ayloffe's Calendar of Welsh and Scotch Rolls, 4th, 1774; Records of Carnarvon, Rec. Comm. 1 vol. fo. 1838
,,   Sessions, Reports on Records of, 1800 Rep. p. 246
* *Wendover, Roger de, Chronicle (Hewlett)*
Western Circuit, Report on Records of, 1800 Rep. p. 242
Westminster School, List of Scholars, 8
Williams', Dr., Library, 79 [Sims, 384]
Wills, 5, 90, 92 [Sims, 343, 349]; Form of Nuncupative, 142-194
,,   Printed, List of some, 91
,,   London, 91
,,   Nuncupative, 91
,,   Prerogative Court, Cant., 91
,,   Vacancy, 91
Window Tax Accounts, Notes on, 1800 Rep. 169
Wines. Account of Wines bought for the King's use, 8 Edw. I., 1837 Rep. 189
,,   Account of Exoneration from Import Duty, 1597-8, 1800 Rep. 172 a
Winton Domesday, 23 [Sims, 6; Cooper, i. 224]
Wood (see Forest), Accounts of Sales of, temp. Hen. VIII. to Edw. VI., 1837 Rep. 13 a
Wools, Documents relating to, Descriptive Slips, Edw. III.—**Record Index, No. 725**
Writs of Election, Parliamentary, 98 [Cooper i. 323]
* *Wyclif. Netter's "Fasciculi Zizaniorum Magistri Johannis Wyclif cum tritico" (Shirley)*

Year Books [Sims, 322; Cooper ii. 391-9]
* *Year Books of Edw. I. and Edw. III. (Harwood and Pike)*
Yorkshire Register Office, 35 [Sims, 37]

A. H. GOOSE AND CO., PRINTERS, RAMPANT HORSE STREET, NORWICH.

Mr. Walter Rye's Publications.

Norfolk Antiquarian Miscellany.

Vol. I. Part 1	-	3 copies for sale	-	price £3	each Part.
2	-	23	,,	-	,, 10s. ,,
II. 1	-	11	,,	-	,, 15s. ,,
2	-	5	,,	-	,, £1 ,,
III. 1	-	17	,,	-	,, 10s. ,,
2	-	19	,,	-	,, 10s. ,,

Calendar of the Feet of Fines for Norfolk.

Part 1	-	54 copies for sale	-	price	15s. each Part.
2	-	54	,,	-	,, 15s. ,,

Monumental Inscriptions in the Hundred of Holt.

33 copies for sale, price 10s. each.

Monumental Inscriptions in the Hundred of Happing.

20 copies for sale, price £1. each.

Three Norfolk Armories.

1 copy for sale, price £2.

Rough Materials for the History of the Hundred of N. Erpingham.

Part 1	-	54 copies for sale	-	price 10s. each Part.
2	-	78	,,	- ,, ,, ,,

Amy Robsart: A Brief for the Prosecution.

Price 1s.

A Month on the Norfolk Broads.

Illustrated by Mr. Wilfrid Ball, 200 copies in stock, price 1s. 6d. nett, post free, 1s. 9d.

Tourists' Guide to Norfolk.

Price 2s.

A History of Norfolk (Popular County Histories Series),

316 pp., price 7s. 6d.; half-bound Roxburgh, 10s. 6d.; large paper, £1. 11s. 6d.

ALL THE ABOVE WORKS CAN BE OBTAINED OF

Messrs. A. H. GOOSE & CO., Antiquarian Printers and Publishers,

RAMPANT HORSE STREET, NORWICH.

DR. MARSHALL'S WORKS.

The Genealogists' Guide,

Being a General Search through Genealogical, Topographical, and Biographical Works, Family Histories, Peerage Claims, &c., by GEORGE W. MARSHALL, LL.D., of the Middle Temple, Barrister-at-Law. Second edition, much enlarged and improved, price 31s. 6d.

Opinions of the Press.

"That Dr. Marshall has conferred an inestimable boon upon genealogical students will be recognised by everyone into whose hands this noble volume falls, and the wonder is how they have done without it so long. The value of the book is simply incalculable, and, while we congratulate Dr. Marshall on its production, we feel bound also to thank him publicly for placing it within the reach of all interested in this class of literary research."—*Notes and Queries.*

"Dr. Marshall's 'Genealogist's Guide' is by far the most complete book of the kind that has hitherto appeared. It is superior to its predecessors, not only in containing many more references, but also—and this is a great matter in a book of this kind—from the fact that what it does contain is arranged in most lucid order."—*Athenæum.*

GEORGE BELL AND SONS, York Street, Covent Garden.

The Genealogist,

Vols. I.—VI.

Edited by GEORGE W. MARSHALL. LL.D., Fellow of the Society of Antiquaries.

Very few complete sets of this book remain for sale. Price £4. 18s. 6d.

GEORGE BELL AND SONS, York Street, Covent Garden.

The Register of Perlethorpe, co. Nottingham,

1528—1812.

Edited by GEORGE W. MARSHALL, LL.D. Privately printed. Price to Subscribers, 21s.

ROBERT WHITE, Bookseller, Worksop.

The Registers of Carburton, co. Nottingham,

1528—1812.

Edited by GEORGE W. MARSHALL, LL.D. Privately printed. Price to Subscribers, 15s.

ROBERT WHITE, Bookseller, Worksop.

www.ingramcontent.com/pod-product-compliance
Lightning Source LLC
Chambersburg PA
CBHW030819270326
41928CB00007B/811